30-SECOND
ECOLOGY

30-SECOND
ECOLOGY

50 KEY CONCEPTS AND
CHALLENGES, EACH EXPLAINED
IN HALF A MINUTE

Editors
Mark Fellowes
Becky Thomas

Contributors
Heather Campbell
James Cook
Julia Cooke
Stephen Murphy
Sarah Papworth
Adam Smith

Illustrator
Nicky Ackland-Snow

IVY PRESS

First published in North America in 2020 by
Ivy Press
An imprint of The Quarto Group
The Old Brewery, 6 Blundell Street
London N7 9BH, United Kingdom
T (0)20 7700 6700
www.QuartoKnows.com

British Library Cataloguing-in-
Publication Data
A catalogue record for this
book is available from the
British Library.

ISBN: 978-0-7112-5965-2

This book was conceived,
designed and produced by
Ivy Press
58 West Street, Brighton BN1 2RA, UK

Publisher **David Breuer**
Editorial Director **Tom Kitch**
Art Director **James Lawrence**
Commissioning Editor **Kate Shanahan**
Project Editor **Elizabeth Clinton**
Design Manager **Anna Stevens**
Designer **Ginny Zeal**
Picture Researcher **Sharon Dortenzio**
Illustrator **Nicky Ackland-Snow**

Cover image: Clipart

Printed in China

10 9 8 7 6 5 4 3 2 1

CONTENTS

INTRODUCTION
Mark Fellowes & Becky Thomas

Humankind's desire to control often has unintended consequences, such as using insecticide to protect crops. Insecticides unintentionally harm pollinators, without which many crops would fail.

All organisms interact with others and with their physical environment. The study of the causes and consequences of these interactions lies at the heart of ecology, which addresses questions about these interactions from a molecular to a global scale. Ecology has always addressed fundamental questions – such as why the tropics are so diverse and why birds lay a certain number of eggs – but today, many critical threats are in immediate need of answering. How do we slow down biodiversity loss, how do we mitigate the environmental effects of climate change, how do we protect people from vector-borne diseases? Ecology works to answer these questions too.

Early ecologists did not have such concerns. Ecology as a science emerged from the great naturalist-explorers of the eighteenth and nineteenth centuries. Luminaries such as Alexander von Humboldt, Charles Darwin and Alfred Russel Wallace travelled the globe, collecting specimens. Their experiences, alongside many others, shed light on the origins of species, the behaviour of these species and the great global patterns of biodiversity. The term 'ecology' (from Greek, meaning the study of where things live) was coined in 1866 by Ernst Haeckel, a German zoologist. By the early twentieth century, ecology had become a hypothesis-driven science, with early work examining topics such as the structure of food webs and the mathematics of predator–prey interactions. From those early beginnings, ecology grew into a fundamental science, but, nevertheless, often with an applied perspective.

In the 1960s, ecology became more radical. Rachel Carson's *Silent Spring* was a clarion call against the threat of environmental degradation, and social action organizations such as Greenpeace and Friends of the Earth brought environmentalism to a worldwide audience. Ecological science underpinned many of their perspectives, but also challenged many assumptions. Ecology began to influence how we viewed the world.

Today we live in the Anthropocene. The influence of humankind on the planet has reached a point where even if we were to all disappear tomorrow, the signs of our presence will remain for millions of years. There is no corner of the Earth that we have not altered. The composition of the atmosphere has been altered by the gasses we release into the air, causing global climate change, acid rain and pollution. We have razed forests for agriculture and forest products, we have removed a vast proportion of fish from the seas and plastics litter the ocean depths, and we have covered the land with concrete for our homes, roads and workplaces. The human population is growing exponentially. Each year, we now add the equivalent of the combined populations of the UK and the Netherlands to the global human population.

The outcomes are habitat degradation and species loss. Many argue that we are in the midst of the sixth great extinction, with the fifth being that which occurred when the dinosaurs were wiped out by a meteor slamming into the Yucatán Peninsula 65 million years ago.

Protecting nature and nature's services will be an increasingly difficult challenge, but there is little choice other than to act now. Ecology helps us do that. Ecology links nature, people and planet in a single subject, seeking to understand the fundamentals of why nature works the way it does, and then taking that understanding and applying it to pressing problems in conservation, habitat management, natural resource use and agriculture. Humans do not exist outside of nature, although insulated homes and air-conditioned offices may suggest otherwise. Humans are an integral part of the global ecosphere, and understanding what that means is more important today than ever before.

This book introduces the main concepts in ecology. Starting with an introduction to **Evolution & Ecology**, it explores how these two principles are closely linked. In many ways, ecology can be thought of as natural selection in action. **Behavioural Ecology** examines how the adaptive behaviour of species affects their ecological interactions. **Population Ecology** is concerned with the population dynamics of species, an outcome of behaviour. Here, we consider how population size is affected by competition for resources – from enemies – and also by the need to disperse. Together, populations form communities, and the chapter **Communities & Landscapes** moves up in scale to examine why ecosystems are so complex, how energy flows through systems and why some species have much more influence on their neighbours than others. Together, communities and the physical environment lead to the formation of biomes, and in the next chapter, **Biomes & Their Biodiversity**, those habitats and species that share common themes and are found around the world are examined – including grasslands, forest and tundra, all of which drive global patterns of biodiversity.

Applied Ecology changes course to consider how this fundamental underpinning can help solve global problems. Here you'll find examples of how ecology can help maintain biodiversity and sustainably feed the world's people. Finally, **Ecology in a Changing World** ends with some of the greatest challenges ecologists currently face, with the emergence of issues such as urbanization and global climate change demanding rapid, radical answers.

Forests are dynamic systems that constantly change through decay and regrowth.

EVOLUTION & ECOLOGY

EVOLUTION & ECOLOGY
GLOSSARY

allopatric speciation The splitting of one species into two due to geographic separation of different populations so that they no longer interbreed and evolve to be different.

asexual reproduction Occurs when offspring arise from a single parent and thus receive genes from only one individual. It is the main form of reproduction in many microbes, such as bacteria, and also occurs alongside sexual reproduction in many plants and fungi.

co-evolution When changes in one species cause natural selection in another species to make a compensatory change and vice versa, for example, in the case of parasite virulence and host resistance.

conspecifics Individuals that belong to the same biological species.

cryptic species Species that look the same to us but can be shown through genetic studies to be different.

DNA barcode A DNA sequence from a particular gene that helps us recognize species because it is very similar in individuals of the same species but differs clearly between individuals of different species.

entomology The study of insects.

eukaryote Organism in which the cells have a nucleus that is enclosed by a membrane and also contain other organelles, such as mitochondria. The eukaryotes are diverse and include single and multicellular organisms, encompassing all plants and animals. Organisms with cells that lack a membrane-bound nucleus are prokaryotes. They are single-celled and include the Bacteria and Archaea.

fecundity The number of offspring that an individual is able to produce.

hybridization Occurs when members of two different species engage in sexual reproduction and produce offspring that carry a mix of genes from both parental species.

lethal selection pressure A selection pressure that kills most members of a population can be a strong evolutionary force if any individuals carry genes that allow them to survive. An important example involves antibiotics that kill most but not all bacteria of a given species.

life history A description of key features of an organism's life, such as size, age at maturity, fecundity and lifespan, and how they are shaped by natural selection.

modern evolutionary synthesis The fusion in the early twentieth century of two key late nineteenth-century insights – natural selection and Mendelian inheritance – with the emerging field of population genetics, to provide a coherent synthesis of how natural selection drives evolution via changes in gene frequencies.

neotropical The neotropics is the tropical region of the Americas.

parataxonomy A term introduced by Daniel Janzen to describe the work of people who sort insect collections into different taxonomic groups that they have learnt to recognize well (such as families of beetles). These sorted specimens may then be inspected by a taxonomist expert in that particular insect group. Parataxonomists are often local people in field study areas and their work can dramatically reduce the workload for expert taxonomists.

pathogen A microbial organism that causes disease in the hosts that it attacks.

polymath A person who has interests and expertise in a wide range of subjects.

speciation The process by which new species are formed when one species splits into two or more.

species description The description and naming of new species; most species on Earth still remain to be formally described and named.

sympatric speciation Speciation that occurs without geographic separation of different populations, i.e. the two new species arise in the same geographical location.

taxonomy The science of describing and naming biological species.

NATURAL SELECTION

the 30-second treatment

3-SECOND SURVEY
Natural selection is the differential survival and reproduction of individuals due to differences in traits that are heritable, having a genetic basis.

3-MINUTE STUDY
Darwin's seminal book, *On the Origin of Species by Means of Natural Selection*, was published in 1859, before the nature of genes was appreciated by scientists. Remarkably, his theory absolutely required a hereditary mechanism, before an appropriate one was known. The dilemma was not solved until the mid-twentieth century, when classical genetics based on Gregor Mendel's research was combined with natural selection theory to yield the 'modern evolutionary synthesis'.

Natural selection is the main force that drives evolution and was proposed in 1858 by Charles Darwin and Alfred Russel Wallace, two outstanding British biologists. It has three key steps. First, individuals of a species vary in many traits. Second, some of these differences influence survival and reproduction. Third, some trait variations are heritable, because they stem from genetic differences. For example, if birds vary in beak length and those with longer beaks survive and reproduce better, then any gene variants (mutations) that increase beak length will be favoured by natural selection and become more common in the next generation. Over generations, this process produces species that appear well-designed (adapted) to their niches. However, no foresight or optimum endpoint is involved. Individuals with the best combination of traits for their current environments will survive and reproduce better than others with less suitable combinations. Consequently, if the environment changes such that birds with shorter beaks now do better, then the gene variants for shorter beaks will be favoured and become more common again. Darwin viewed natural selection as a gradual process, but it can be very fast. For example, lethal selection and a very short generation time can combine to drive rapid evolution of antibiotic resistance in bacteria.

RELATED TOPICS
See also
MAKING NEW SPECIES
page 16

LIFE HISTORY TRADE-OFFS
page 24

3-SECOND BIOGRAPHIES
CHARLES DARWIN
1809–82
Naturalist and biologist who laid out the detailed case for evolution by natural selection in *The Origin of Species*.

GREGOR MENDEL
1822–84
Silesian monk and scientist whose experiments provided the foundation for classical genetics.

ALFRED RUSSEL WALLACE
1823–1913
Naturalist, explorer and polymath who conceived the idea of natural selection independently from Darwin.

30-SECOND TEXT
James Cook

The origin of new species by evolution from a common ancestor was a controversial idea.

MAKING
NEW SPECIES

the 30-second treatment

New species arise by one species splitting into two, but how? Ernst Mayr was a very influential thinker on this topic and championed the role of geographic isolation. Many bird species have a mainland population but a different variety, or closely related species, on an offshore island. If there is little or no exchange of breeding individuals between island and mainland, then each population can evolve and adapt to its own local conditions until it becomes a separate species. Such 'allopatric' speciation does not require islands – other geographic barriers like mountains and rivers can suffice – and is thought to be widespread. In contrast, it has long been controversial whether much speciation happens without geographic isolation. Such 'sympatric' speciation is easily envisaged, however, with insects that only feed and mate on one type of host plant. If some individuals colonize a new host plant species and show fidelity to this, then the two insect populations are reproductively isolated and can speciate, even if feeding on plants in the same field. The sympatric speciation debate has waned recently, as researchers increasingly ask which ecological factors – such as resource use, mate choice and even parasites – drive speciation. In addition, advances in genomics are allowing scientists to study how many and which genes may be involved in speciation.

3-SECOND SURVEY
Speciation requires populations to diverge genetically until they cannot interbreed; often this follows geographic isolation, but speciation can also occur locally, driven by ecology.

3-MINUTE STUDY
Most thinking on speciation centres on the problem of how two species come from one. However, an alternative is to make three species from two, when two 'parent species' interbreed to make a new species through hybridization. In animals, hybrid individuals tend to have low viability or be sterile, like a mule. However, hybridization seems to be an important source of new species in plants.

RELATED TOPICS
See also
NATURAL SELECTION
page 14

COUNTING SPECIES
page 18

3-SECOND BIOGRAPHY
ERNST MAYR
1904–2005
German-American evolutionary biologist who pioneered conceptualization of the nature of species and speciation processes.

30-SECOND TEXT
James Cook

Reproductive isolation helps populations split into new species. Oceanic islands and host islands can both provide this.

COUNTING SPECIES

the 30-second treatment

Before species are counted,

they must first be found and described by a taxonomist, who checks they are different from other species. Carl Linnaeus was the father of modern taxonomy, and the current count stands at around 1.3 million named species. This figure includes about 1 million animals, 200,000 plants, and 100,000 other organisms, such as fungi. But there are lots more unnamed species, especially small invertebrates. To estimate the real totals, numbers are extrapolated from known species diversity. For example, to estimate the global diversity of beetles, beetles could be sampled widely across regions and habitats. If 50 per cent of species collected were previously unknown, then the real number is about double the current number described. This is a simple example, and there are other methods to count 'missing species', but they all use similar logical approaches to estimate the real number from known figures. Recent studies like this suggest that there are actually about 9 million eukaryote species, including 7.7 million animals (mostly insects) and 600,000 fungi. Unfortunately, it is far harder to count species of bacteria, because many different species look much the same. In addition, the species concepts that we use for plants and animals don't work well with organisms that reproduce asexually (clonally), such as bacteria.

3-SECOND SURVEY
There could be about 9 million eukaryote species, but numbers are more certain for some groups (10,000 birds) than others (5 million insects).

3-MINUTE STUDY
Taxonomists identify known species and describe new ones, but some species look so similar that even taxonomists cannot tell them apart. However, such 'cryptic species' can now be told apart using their DNA sequences. If cryptic species turn out to be common in insects, for example, then the global species count will need revising upwards. As many microbes look very similar, biologists already rely heavily on DNA barcodes to recognize and count bacterial and fungal species.

RELATED TOPICS
See also
NATURAL SELECTION
page 14

MAKING NEW SPECIES
page 16

3-SECOND BIOGRAPHY
CARL LINNAEUS
1707–78
Swedish physician, botanist and zoologist who was the father of modern taxonomy and introduced the binomial naming system for species.

30-SECOND TEXT
James Cook

Carl Linnaeus started the formidable task of naming the millions of species on Earth.

18 January 1939
Born in Milwaukee,
Wisconsin, USA

1952
Takes formative trip to
Mexico with family

1961
Completes BSc in Biology

1965
Awarded PhD
from Berkeley

1966
Publishes paper on
co-evolution of ants
and acacias

1970
Proposes Janzen–Connell
hypothesis

1976
Joins faculty at
the University
of Pennsylvania

1986
Proposes protection
for Guanacaste forest

1987
Listed on UNEP Global
500 Roll of Honour

1989
Guanacaste National
Park established

1992
Elected to National
Academy of Sciences,
USA

1996
Appointed Endowed
Chair, University of
Pennsylvania

1997
Founds Guanacaste
Conservation Fund

1997
Awarded Kyoto Prize
(Basic Sciences Field)

2004
Reveals '10 species in
one' with DNA barcodes

2014
Wins Blue Planet Prize

DANIEL H. JANZEN

Daniel Hunt Janzen was born in Milwaukee, Wisconsin, and grew up in Minnesota in northern USA. However, it was on a long family road trip around Mexico, collecting butterflies, that the 14-year-old naturalist developed his fascination with tropical biodiversity. He subsequently completed a BSc in Biology at the University of Minnesota, and eagerly took every opportunity to immerse himself in neotropical fieldwork.

Janzen's PhD programme at the University of California, Berkeley, allowed him to pursue research in Mexico, and also experience fieldwork at several sites in Costa Rica, a country with which he is now almost synonymous. His PhD was awarded in 1965 and focused on the intriguing interactions between acacia trees and the pugnacious ants that live within their thorns and protect the tree from herbivores. His PhD was pioneering in its use of field experiments and fusion of botany and entomology. It also kick-started the wider study of co-evolution between interacting species.

In 1970, Janzen and ecologist Joseph Connell independently came up with what is now known as the Janzen–Connell hypothesis. This suggests that the high tree diversity in tropical forests is a result of host-specific natural enemies (such as herbivores and pathogens). These kill off seedlings growing near conspecifics, but other tree species and those further away escape attack, promoting a forest with a patchwork of many different tree species.

In 1976, Janzen joined the University of Pennsylvania, which remains his academic home to this day. Meanwhile, growing concern for the future of tropical biodiversity led Janzen and Winnie Hallwachs, his wife and research partner, towards conservation action. In Costa Rica's Guanacaste forest, they brought together detailed understanding of forest ecology, local human societal interests and wildland restoration. Their call for protected status for Guanacaste in 1986 was answered in 1989 with the establishment of a National Park and wider Conservation Area.

Another long-term project involves the discovery and description of Costa Rica's biodiversity. In this, Janzen champions the role of parataxonomists – local people without university training but with excellent skills and understanding of local nature. The research has revealed widespread host specialization by natural enemies, as envisaged in the Janzen–Connell hypothesis. Since around 2000, this work has also embraced DNA barcoding as a rapid way to make sense of the huge diversity of undescribed and cryptic tropical insect species.

James Cook

HABITATS & NICHES

the 30-second treatment

3-SECOND SURVEY
A niche describes the set of conditions under which a given species can live, and what it does there.

3-MINUTE STUDY
The Hutchinsonian niche is conceptualized as an 'n-dimensional hypervolume' where dimensions represent environmental conditions or resources. Hutchinson wondered how so many different species could live in one habitat and inspired theoretical modelling of this issue. The notion that any two competing species must differ along at least one niche dimension links strongly to the competitive exclusion principle – if they don't differ, the stronger will eventually eliminate the weaker competitor.

The ecological niche represents the set of physical and biological conditions in which a given organism or species can prosper. The first step to seeing a particular organism is to look in the right habitat. Lakes, grasslands and woodlands can occur close together, but they are home to different sets of species, many of which may be restricted to only one habitat. Further, within one habitat, there are many species, each with its own particular niche. Two different types of niche are commonly distinguished. The fundamental niche focuses on the physical environment (such as climate conditions) and habitat that can sustain a given species. However, species also face biotic challenges – competitors, predators and parasites – and biological interactions with these enemies tend to restrict the species to realized niches that are smaller than their fundamental niches. Another key issue is dispersal. The fundamental niche of a European grassland plant might include similar habitats in Australia and Argentina, but natural dispersal over such distances is unlikely. However, humans often provide the dispersal needed for species to expand their realized niches, too often leading to problems with invasive species that have left their enemies at home.

RELATED TOPICS
See also
DISPERSAL
page 50

COMPETITION
page 66

3-SECOND BIOGRAPHY
GEORGE EVELYN HUTCHINSON
1903–91
British zoologist and ecologist who played a key role in developing ecological niche theory and driving research on competition between species.

30-SECOND TEXT
James Cook

Each species has its own ecological niche, so don't expect to see it elsewhere!

LIFE HISTORY TRADE-OFFS

the 30-second treatment

Trade-offs are a familiar part of most people's lives. Should you go out for dinner or save your money for a holiday? Do you watch that new film or have a game of tennis? What these examples share is a choice between alternative uses for a limited resource (here, money or time). The same concept is important for species' life histories, since biological resources are generally limiting and an organism must allocate them across key activities, such as growth, survival and reproduction. Evolutionary theory explores which circumstances favour particular life histories and this depends on understanding trade-offs between traits. A fundamental trade-off is between growth and maintenance versus reproduction. How long should an organism focus on growth before switching to reproduction? Species vary greatly, and both models and comparative studies show that high mortality and an unpredictable environment favour early reproduction and high fecundity. This is intuitive since investment in your own growth and maintenance may well be wasted in a risky environment. Population studies provide direct evidence for the general trade-off between survival and reproduction. For example, in some songbird species, parents that rear many offspring in the breeding season are more likely to die in the next winter season.

RELATED TOPICS
See also
NATURAL SELECTION
page 14

HABITATS & NICHES
page 22

3-SECOND BIOGRAPHY
RICHARD LAW
1950–
British ecologist and evolutionary biologist who introduced the thought experiment of the impossible Darwinian Demon, which is great at everything in its life history.

30-SECOND TEXT
James Cook

Growth versus reproduction is a classic life history trade off.

BEHAVIOURAL ECOLOGY

BEHAVIOURAL ECOLOGY
GLOSSARY

altruistic behaviour When an individual behaves in a way that benefits others but has a cost for the individual doing the behaviour.

echolocation Using sounds to navigate. Used by some bats, dolphins and also birds and shrews.

evasion strategies Behaviours used by animals to avoid being caught by predators.

inclusive fitness A measure of an individual's success in passing on genes, including direct offspring (direct fitness) and the offspring of relatives (indirect fitness), where the individual's behaviour has contributed to their survival.

intersexual selection When one sex is attracted to mates of the other sex by a particular characteristic, and chooses their mate. This is also called mate choice.

intrasexual selection When individuals compete with others of the same sex for access to mates.

lactation Female mammals produce milk to feed their young. This process, and the period when the female produces this milk, is called lactation.

migration Large-scale movements of animals, usually cyclical movements between specific locations that are used at different times. Many migrations happen at predictable times, such as birds moving from tropical regions to colder regions in spring, and returning in autumn.

monogamy When males and females both have a single mate in a breeding season, forming a pair.

polyandry When females have many mates in a breeding season.

polygyny When males have many mates in a breeding season.

promiscuity When both males and females have many mates during a breeding season.

relatedness Describes how closely related two individuals are, measured using the coefficient of relationship. The coefficient of relationship varies between 100 per cent (individuals with identical genes, such as identical twins) and close to 0 per cent (individuals that are very distantly related and have few genes in common). A parent and child have a coefficient of relationship of 50 per cent, as on average they will have 50 per cent of their genes in common.

selfish herd A group of individuals collecting with others to take advantage of dilution in predation risk. All individuals in the group compete to gain the safest position within the group.

simian immunodeficiency virus A disease found in African monkeys and apes caused by a retrovirus that is closely related to the retrovirus that causes human immunodeficiency virus (HIV). When infected with simian immunodeficiency virus, these African species do not seem to suffer any negative symptoms.

SEXUAL SELECTION

the 30-second treatment

3-SECOND SURVEY
Characteristics that increase chances of mating success are selected for over time, leading to elaborate ornaments or weapons.

3-MINUTE STUDY
Although most examples of sexual selection are found in male animals, sexual selection also occurs through male choice and female–female competition, leading to elaboration of female characteristics. This most often happens when males provide parental care. In northern jacanas, a wading bird in central America, males incubate the eggs, whereas female jacanas have longer wing spurs, which are used in aggressive interactions with other individuals.

Many animals have physical features that are not necessary for survival, such as the elaborate plumage of male birds of paradise, or the large antlers of male deer. When these features increase an individual's chances of mating successfully, they can become increasingly extravagant over many generations, in a process called sexual selection. Sexual selection can arise through mate choice (also called intersexual selection), where a particular characteristic, such as the number of 'eyes' on a peacock's ornate tail, will attract mates and allow an individual to have more offspring. These features sometimes cost energy for individuals to produce, so are an honest signal of good condition to potential mates. When choosing a mate, both male and female blue-footed boobies will lift and display their feet, which are less blue when they are deprived of food. Sexual selection can also occur through intrasexual selection, where individuals of the same sex compete for mates. Silverback male mountain gorillas will beat their chests with their fists, stamp their feet and charge other males to defend their harem of females. Sometimes fights escalate and the intruder will win, becoming the new male of the group. When he mates with the females, he will pass winning features, such as large body size, to his offspring.

RELATED TOPICS
See also
LIFE HISTORY TRADE-OFFS
page 24

MATING SYSTEMS
page 32

COSTS OF REPRODUCTION
page 34

3-SECOND BIOGRAPHIES
CHARLES DARWIN
1809–82
Proposed the theory of sexual selection to explain impractical male decorations such as peacock tails.

NANCY T. BURLEY
fl. 1977–
Studies sexual selection by adding artificial ornaments to the heads and legs of birds.

30-SECOND TEXT
Sarah Papworth

Animals have evolved elaborate features to gain mates, such as blue feet or large antlers.

MATING SYSTEMS

the 30-second treatment

3-SECOND SURVEY
Males and females need a mate to reproduce, but some search for multiple mates or a mate to help care for young.

3-MINUTE STUDY
Some animals form groups where many adults care for the young of dominant individuals. Up to 50 meerkats live in a group, but four of every five young are produced by the dominant female. When a dominant meerkat is pregnant, she harasses other females in the group and temporarily evicts them to prevent them from reproducing. The evicted females are allowed to return and care for the young after the dominant female has given birth.

All animals that reproduce

sexually need to find a mate. Mating systems describe how animals find mates and how many mates they have. The type of mating system an animal has is influenced by how easy it is to find mates and how much parental care its young will need. Monogamy is when a male and female mate exclusively. The wandering albatross mates for life, so new pairs spend more than two years establishing a bond before breeding. Monogamy is common in birds such as the wandering albatross because two parents are required to successfully incubate eggs and care for chicks. Polygyny is when males have many mates in a breeding season. A single male red deer will defend a harem of many females. The opposite is polyandry, for example, queen honey bees, which mate with multiple males. Polygamy is a general term that describes both polygynous and polyandrous mating systems. The final mating system is promiscuity, where both males and females have many mates during a breeding season, such as chimpanzees. Some animals seem monogamous as they form pairs to care for offspring, but both the male and female will try to mate with others. In tree swallows, half of all nests contain young that are unrelated to the male providing food.

RELATED TOPICS
See also
SEXUAL SELECTION
page 30

GROUP LIVING
page 38

3-SECOND BIOGRAPHY
LEWIS ORING
fl. 1962–
Studies the polyandrous spotted sandpiper and in 1977 wrote a classic review of mating systems with Stephen Emlen.

30-SECOND TEXT
Sarah Papworth

Wandering albatross pairs bring food to their chick, which can't fly until it is around ten months old.

COSTS OF REPRODUCTION

the 30-second treatment

The ocean sunfish produces millions of eggs when it spawns, but other animals, such as elephants, produce only a few calves in a lifetime. Elephants care for their young, defending them from attack, while sunfish fry are left to fend for themselves, and many die. The costs and benefits of reproduction influence how much an individual will invest in each of their offspring, which then determines how many offspring they have. Each spring, pairs of great tits lay 5 to 12 eggs. After hatching, both parents spend all day feeding the chicks, but how much food they can bring is limited by how many caterpillars are available. When there are more chicks, each has less to eat and is less likely to survive the winter, but in a good caterpillar year, all may survive. The effort of caring for more offspring can also affect the survival of parents. When female Townsend's voles reproduce before they reach adult size and have more than five young in a litter, both the mother and young are unlikely to survive the three-week lactation period. These females do not have enough energy reserves to continue growing to adult size and produce milk for five offspring. In contrast, females that don't reproduce until they are adult size can successfully care for multiple litters of eight offspring.

3-SECOND SURVEY
Plants and animals are unable to have unlimited offspring as resources are limited and reproduction has costs for parents.

3-MINUTE STUDY
The costs of reproduction not only affect the mother. Cotton-top tamarins are small monkeys that usually give birth to twins. Although the infants depend on their mothers for milk, other members of the group will carry them between feeds. The father and older male siblings will spend more time carrying infants than females. Individuals that carry infants spend less time eating and can lose more than 10 per cent of their body weight in a single week.

RELATED TOPICS
See also
LIFE HISTORY TRADE-OFFS
page 24

SEXUAL SELECTION
page 30

MATING SYSTEMS
page 32

3-SECOND BIOGRAPHY
DAVID LACK
1910–73
British evolutionary biologist who started a study of reproduction by great tits near Oxford, England, in 1947, which is still running today.

30-SECOND TEXT
Sarah Papworth

Searching for food is time-consuming, but it's easier to feed many chicks if there are lots of caterpillars about.

DEFENCE AND COUNTER-DEFENCE

the 30-second treatment

Predators need to eat, but prey

will try to avoid being eaten. Prey animals are more likely to survive if they have characteristics that reduce their chances of predation, such as leaf insects that masquerade as leaves, making them less obvious to predators. Predators thrive if they can overcome these defences, such as domestic chicks that can learn to identify caterpillars that look like twigs. Over time, this conflict between predators and prey can lead to 'arms races' where the strategy of each group constantly changes to outcompete the other. Bats use echolocation to find their prey, but some moths can detect echolocation calls and adopt evasion strategies. Some bats, in turn, now echolocate at frequencies higher or lower than these moths can hear. Similar conflicts can be observed between parasites and their hosts. Common cuckoos are brood parasites, meaning females lay their eggs in the nests of unwitting hosts, who will raise the young cuckoo if they don't detect the cuckoo egg. In response, host birds such as the pied wagtail have developed highly individual egg markings, allowing them to identify and reject over 90 per cent of cuckoo eggs. But common cuckoo eggs can also mimic the appearance of host eggs, and each cuckoo lineage now lays eggs that mimic those of their preferred hosts.

3-SECOND SURVEY
Prey adopt defences to escape predators, but predators will overcome these, so prey develop counter-defences in a constant cycle.

3-MINUTE STUDY
Pit-viper venom contains up to 40 different toxins, which pit vipers use to immobilize and kill their prey. In the Americas, some opossums produce proteins that neutralize pit viper toxins. The genes that produce these proteins are inherited, allowing certain opossum lineages to capture and eat pit vipers. However, the pit vipers also eat opossums, making it unclear whether opossum proteins are a predator adaptation or prey defence.

RELATED TOPIC
See also
GROUP LIVING
page 38

3-SECOND BIOGRAPHIES
LEIGH VAN VALEN
1935–2010
Proposed the Red Queen hypothesis to describe the constant adaptations of predators and prey.

CLAIRE SPOTTISWOODE
fl. 1999
Showed that prinia birds use egg colour and pattern to identify and reject cuckoo-finch eggs.

30-SECOND TEXT
Sarah Papworth

Moths that can detect the calls of their bat predators will move away, and those that can't will be caught.

GROUP LIVING

the 30-second treatment

Some groups are short-term associations of individuals, such as wildebeest travelling together during annual migrations. However, other animals live in permanent and relatively stable groups with complex social systems. Group living can be explained by examining the costs and benefits for individuals, particularly due to predation and access to resources. Predators can be less able to focus on a single individual in larger groups, and for each individual prey the chance of being attacked reduces. Anti-predator behaviour by other group members also benefits individuals. In larger groups, each ostrich spends more time feeding and less time with its head up scanning, but total group vigilance increases as there is more time when at least one individual is scanning. Living in groups can also provide individuals with information about where to find food, or better opportunities to capture food. Some Antarctic killer-whale groups coordinate their movements to swim rapidly towards inaccessible seals on ice floes. The wave generated washes the seal into the sea. The whales capture and share the seal, which is a potential cost of group living: individuals have to compete with other group members. Therefore, whether animals live in groups, and how big those groups are, will depend on local conditions, such as the availability of food and density of predators.

RELATED TOPICS
See also
MATING SYSTEMS
page 32

DEFENCE AND
COUNTER-DEFENCE
page 36

WILLIAM D. HAMILTON
page 42

3-SECOND BIOGRAPHIES
WILLIAM D. HAMILTON
1936–2000
English evolutionary biologist who suggested the 'selfish herd' composed of individuals collecting with others to take advantage of dilution.

IAIN D. COUZIN
fl. 1995
British scientist who uses computer simulations to describe why animals form groups and how they make collective decisions.

30-SECOND TEXT
Sarah Papworth

3-SECOND SURVEY
Individuals live in groups for foraging and anti-predator benefits but can incur costs from exposure to disease and competition with other group members.

3-MINUTE STUDY
Living in groups can have costs for individuals. Predators may be more likely to detect and attack large groups. Groups of 25 redshank, a shorebird, are twice as likely to be attacked by sparrowhawks, compared to groups of fewer than 10. Diseases and parasites can spread more quickly in large groups. Cliff swallows group in nesting colonies, and larger colonies have more blood-sucking swallow bugs in each nest.

Living in groups means there are more individuals to look out for danger, so predators may be spotted sooner.

KIN SELECTION
the 30-second treatment

Kin selection theory was

developed to explain seemingly altruistic animal behaviours that are costly to individuals but which benefit their relatives. For example, ground squirrels are more likely to give alarm calls to predators when they have close relatives nearby, even though calling increases the risk of being attacked by the predator. The spread of genes that generate altruistic behaviour depends on the reproductive cost to an individual, and how closely related the individual is to those who benefit. Individuals share approximately half their genes with full siblings and parents, and one eighth of their genes with first cousins. This measure of relatedness is used in Hamilton's rule, which states altruistic behaviour should occur if reproductive cost to the altruistic individual is outweighed by the reproductive benefit to a relative, multiplied by the degree of relatedness between the two individuals. An extreme example is found in some species of social insects such as honey bees, where most individuals are sterile females that spend their lives helping to raise the offspring of the queen. Although the cost to sterile individuals is high, all the individuals in the colony are the offspring of the queen, so sterile individuals are ensuring their own genes survive by helping to raise their sisters and brothers.

RELATED TOPICS
See also
SEXUAL SELECTION
page 30

GROUP LIVING
page 38

WILLIAM D. HAMILTON
page 42

3-SECOND SURVEY
'I would lay down my life for eight cousins or two brothers' – J. B. S. Haldane.

3-MINUTE STUDY
Kin selection requires individuals to interact with relatives, which may be achieved through kin recognition. When social amoebae are starving they group together. One in five of the amoebae form a stalk and die, while the rest are released as spores to find areas with more food. Individuals tend to form these fruiting bodies with very close genetic relatives. Two genes that vary between different lineages of social amoebae allow kin to recognize each other.

3-SECOND BIOGRAPHY
JOAN STRAUSSMAN
fl. 1979–
Uses genetic relatedness to understand kin selection in social insects and social amoebae.

30-SECOND TEXT
Sarah Papworth

Most bees in a hive are female and all are closely related, but only the queen bee produces eggs.

1 August 1936
Born in Cairo, Egypt

1939
Temporarily evacuated
from Kent to Edinburgh
during World War II

1960
Graduates from the
University of Cambridge
with a degree in Genetics

1964
Starts teaching at
Imperial College London

1964
'The Genetical Evolution
of Social Behaviour'
paper published

1968
Earns a PhD from
University College
London and the London
School of Economics

1978
Starts working at the
University of Michigan,
Ann Arbor

1980
Elected Fellow of the
Royal Society of London

1981
'The Evolution of
Cooperation' paper
published

1982
Hamilton-Zuk hypothesis
on parasites and sexual
selection published

1984
Returns to England as
Research Professor at the
University of Oxford

1988
Awarded the Darwin
Medal of the Royal
Society

7 March 2000
Dies in London, England

WILLIAM D. HAMILTON

Many of the underlying theories in behavioural and evolutionary ecology were proposed by William Donald Hamilton. Hamilton was born in Cairo in 1936, while his father, Archibald Hamilton, was briefly posted there with the British Corps of Royal Engineers. Although both Hamilton's parents were New Zealanders, he spent most of his childhood in rural Kent, England, where he grew up fascinated by insects and evolution.

Hamilton went to Tonbridge School and was awarded a state scholarship to study Genetics at the University of Cambridge. He felt there was not enough focus on natural selection at Cambridge, but he did meet and work with Professor Ronald Fisher, whose book *The Genetical Theory of Natural Selection* inspired much of his later work.

While studying for his PhD in Mathematical Genetics at the London School of Economics in 1964, Hamilton published the paper 'The Genetical Evolution of Social Behaviour'. He introduced the concept of inclusive fitness, which describes how an individual's behaviour can change both their own and their relative's reproductive success. Using social insects as an example, he applied this idea to explain the evolution of altruistic behaviour, which causes costs for individuals but is beneficial to relatives. The relationship between relatedness and the costs and benefits of altruism are now known as Hamilton's rule. Inclusive fitness and Hamilton's rule did not receive much attention initially, but now are fundamental concepts in kin selection. Later in his career, Hamilton promoted the work of scientists he felt were undervalued, arguing his early work would have been recognized more quickly if championed by a senior scientist.

After early work on social behaviour, Hamilton worked on sex and sexual selection from the 1980s. With the behavioural ecologist Marlene Zuk, he developed the Hamilton–Zuk hypothesis of parasites and sexual selection, which suggests that choosy mates prefer visible characteristics, which reliably indicate genes for parasite resistance. Although originally proposed for animal systems, the hypothesis inspired research on human mate choice.

Throughout his career Hamilton conducted fieldwork, from a Brazilian trip to study social insects in 1963, to a trip to the Congo to study simian immunodeficiency virus shortly before his death in 2000. He also had influence outside of ecology, with a paper on 'The Evolution of Cooperation' co-authored with Robert Axelrod in 1981, which has become a hugely influential paper in political science. Hamilton's combination of original ideas, population genetics models and naturalist knowledge created many influential papers, which were compiled and published with autobiographical comments as *Narrow Roads of Gene Land* (OUP, 1998).

Sarah Papworth

POPULATION ECOLOGY

allelopathy The inhibition of germination or growth of one organism by another, through the release of chemical substances into the environment

aposematism The conspicuous colours or patterns on an animal that advertise to potential predators that it tastes bad or is toxic.

co-evolution The in-tandem evolution of two or more species as a result of selection pressures exerted by each on the other.

gene flow The transfer of genetic variation from one population to another.

genus A taxonomic category ranking above species but below family.

germinate To begin to grow after dormancy, particularly referring to a seed or spore.

hemiparasite Plant that obtains part of its food (fixed carbon) by parasitism, but is also capable of photosynthesis (carbon fixation).

holoparasite Plant that obtains all of its food (fixed carbon) from a host plant and does not photosynthesize.

inbreeding Breeding of individuals that are closely related genetically.

polyp One of two body forms of organism in the phylum Cnidaria. Swimming medusae are the other form, which differ from the sessile polyps.

population cycle A phenomenon where populations of a species rise and fall over predictable time periods.

predation The preying of one animal on others.

POPULATION DISTRIBUTIONS

the 30-second treatment

Population distribution describes how individuals within a population are scattered across the area they occupy. Individuals in some populations disperse uniformly across an area, such as nesting gannets, which ensures maximum distance between birds. Uniform distributions can be driven by social interactions and/or competition. In other populations, organisms are clumped, such as herds of animals that group together for safety or social benefits. Clumped distributions also occur when individuals congregate around patchy resources. Species are often best adapted to the central areas of their distribution, with conditions less favourable on the fringes forming the limits of the population; individuals are less densely distributed at these fringes. The human population is not even across the world, with some countries more densely populated than others and uneven population distributions within countries. In Australia, over 80 per cent of the population lives within 50 km (30 miles) of the coast, despite the large size of the country. The distribution of individuals in a population can also be random. This is comparatively rare, as it requires consistent environmental conditions and resources, and a lack of strong interactions among individuals. Random distribution is most common in species such as dandelions that rely on wind and water for dispersal.

RELATED TOPICS
See also
DISPERSAL
page 50

LIMITS TO POPULATION
GROWTH
page 52

METAPOPULATIONS
page 58

3-SECOND SURVEY
Individuals in populations are often clumped around resources or gather together for group benefits, but uniform and random population distributions exist too.

3-MINUTE STUDY
Some species of plant have populations with individuals distributed evenly where the pattern is driven by the plants themselves. Purple sage (*Salvia leucophylla*) in the southwest of the US produces chemicals that inhibit the growth of other sage plants around it, causing the uniform distribution of individuals. This process is known as allelopathy, and the chemicals some plants produce affect only others of the same species, while others deter all other plants.

3-SECOND BIOGRAPHY
MARIE-JOSÉE FORTIN
1958–
Canadian ecologist whose work shows how changes in land use affect the persistence of populations.

30-SECOND TEXT
Julia Cooke

Human populations often live close together for social benefits. The largest cities in Australia are coastal, where conditions are most favourable.

DISPERSAL

the 30-second treatment

3-SECOND SURVEY

Dispersal is the movement of living things from the site of their birth to their breeding site; that is, movement leading to gene flow.

3-MINUTE STUDY

Organisms have differing dispersal strategies because a parent can allocate resources between many or few offspring. Producing many small offspring means more chances for some making it to adulthood, but at a lower likelihood of individual survival. Corals produce millions of polyps: most will be eaten or land in hostile locations, but some will survive to breed. Elephants give birth to few, well-developed young and care for them for years before they move away.

To be successful, organisms

must survive from birth to reproduction, but remaining at a birthplace can mean competition with parents and siblings for food and/or space, or can expose individuals to higher levels of attack by predators or diseases. Dispersal helps reduce these risks. Successful dispersal is the movement that results in finding suitable habitats, food and reproductive partners, and it also can result in a species expanding its range. Many animals and micro-organisms can fly, swim or crawl away from their parents. Seeds have diverse shapes to help dispersal – to fly in the wind like helicopters, float on water, be swallowed or attach to fur to catch a ride from passing animals. Scientists have tracked seed dispersal and recorded a 'dispersal tail' of a small percentage of seeds that travels much further than others; most seeds fall close to the parent, fewer travel further away and very few might travel far away from their origin. Dispersal involves three distinct stages – departure, transfer and settlement – with challenges at each stage. For example, a fungal spore could be damaged while still enclosed in a puffball, be eaten while travelling on a breeze, or land somewhere unsuitable for it to grow. Dispersal is an essential but challenging part of life.

RELATED TOPIC

See also
MATING SYSTEMS
page 32

30-SECOND TEXT

Julia Cooke

To spread offspring away from their parents, species have evolved ways of dispersing, using wind, water and their own movement.

LIMITS TO POPULATION GROWTH

the 30-second treatment

3-SECOND SURVEY

Parents can produce many offspring, meaning that populations grow; but with limited resources, population growth is controlled as individuals compete and disease spreads more easily.

3-MINUTE STUDY

Human activity affects the population growth of many species. The sex of crocodiles is determined by the temperature in which eggs are incubated, with hot air temperatures and cooling rainfall driving how many male versus female crocodiles emerge each year. With rising global temperatures and less regular rainfall, some crocodile populations are increasingly male dominated, which reduces population growth as there are fewer females to lay eggs. This could lead to the extinction of some populations.

Typically, parents produce more than one or two offspring, which means they breed beyond just replacing themselves, and so populations grow. But populations cannot develop indefinitely as there is a limit to the resources they use or need. Populations can be limited by space or food, with individuals competing for food, breeding sites, water or any other resource needed to complete their life cycle. Such competition limits population growth rates by lowering birth rates, increasing death rates and driving emigration. An example is seen in parrot species in Australia, many of which nest in tree hollows, and so their population growth is limited by the amount of tree hollows available. The size of a population that an area can support is described as the carrying capacity. Some limits to population growth are dependent on the density of the population. Denser populations allow parasites and diseases to spread more easily, and help predators to find and capture food, which can in turn limit population growth. For example, when Atlantic salmon occur in high numbers, sea lice, which attach to fish and feed on their skin, are spread more readily among the salmon. Other population limits are independent of density. For example, weather events, fires and floods can limit the population growth of species irrespective of how dense their population.

RELATED TOPICS

See also
CYCLING POPULATIONS
page 56

PARASITES
page 64

COMPETITION
page 66

3-SECOND BIOGRAPHY
PAUL R. EHRLICH
1932–
American biologist who has published books on the predicted limits to human population growth and the associated environmental impacts.

30-SECOND TEXT
Julia Cooke

Crocodile populations may be limited by global temperature increases while other species can be limited by parasites.

29 May 1932
Born in Philadelphia,
Pennsylvania, USA

1953
Gains BA degree,
University of
Pennsylvania

1954
Marries Anne Howland,
a population biology
researcher

1955 & 1957
Gains Masters degree
and completes PhD,
University of Kansas

1957
Becomes Research
Associate, University
of Kansas

1959 & 1962
Appointed Assistant
Professor, and later
Associate Professor,
of Biological Sciences,
Stanford University

1963
Publishes *The Process
of Evolution*

1966–
Becomes Professor
of Biological Sciences,
Stanford University

1968
Publishes controversial
book *The Population
Bomb*

1977
Appointed Bing Professor
of Population Studies,
Stanford University

1987
Awarded Gold Medal,
WWF International

1990
Wins Crafoord Prize
in Population Biology
and Conservation of
Biological Diversity

1994
Awarded UN Environment
Programme Sasakawa
Environment Prize

1999
Wins Blue Planet Prize

2001
Awarded Eminent
Ecologist Award,
Ecological Society
of America

2008
Publishes *The Dominant
Animal*

2014
Gains Honorary
Fellowship, Royal
Entomological Society
of London

2016
Appointed Emeritus
Professor of Biological
Sciences and Emeritus
Bing Professor of
Population Studies,
Stanford University

PAUL R. EHRLICH

Paul Ralph Ehrlich was born in Philadelphia, USA, in 1932 and spent part of his childhood in New Jersey. His early interest in ecology was in part inspired by *Road to Survival* by William Vogt. After completing his degree at the University of Pennsylvania, and Masters and PhD research at the University of Kansas, Ehrlich studied the genetics and behaviour of parasitic mites. Ehrlich joined Stanford University in 1959 as an assistant professor, becoming full professor in 1966. He is currently the Bing Professor of Population Studies, Emeritus, at Stanford and resides in California with his wife, Anne Ehrlich.

Ehrlich conducted long-term studies on the population dynamics, structure and genetics of butterflies, which have been applied to both controlling and preserving populations. His fieldwork took place in North and South America, Africa, Asia and Australia, with field studies in his home country largely based in California and Colorado. With botanist and environmentalist Peter Raven, he co-founded the field of co-evolution, using examples from plant–herbivore interactions, and he wrote the widely used textbook *Process of Evolution* with Stanford biologist Richard W. Holm.

Ehrlich is a pioneer of forewarning the public about problems of overpopulation, especially pressures on non-renewable resources. Among his many books, *The Population Bomb* and *The Dominant Animal*, written with Anne Ehrlich, have influenced thinking about the negative impacts of humans on Earth. Ehrlich highlighted the increasing rate of population growth: the human population reached 1 billion people as recently as 1850, but it only took another 80 years, until 1930, to reach 2 billion. His work raised concerns about the resultant pressure on natural resources, which he put in perspective by indicating that the magnitude of anthropological processes was approaching and could eclipse equivalent natural processes (for example, weathering versus mining rates). His predictions about population collapse and large-scale environmental change have been criticized, but Ehrlich has argued that using his ideas to avoid predicted catastrophes would lead to a healthier society, irrespective of any projection inaccuracies.

Ehrlich is interested in the human cultural evolution of environmental ethics and is involved in both conservation and policy initiatives, including rehabilitating human-disturbed landscapes. Ehrlich is the recipient of many awards, including the Craford Prize of the Royal Swedish Academy of Sciences, United Nations' Sasakawa Environment Prize and the Eminent Ecologist Award of the Ecological Society of America.

Julia Cooke

CYCLING POPULATIONS

the 30-second treatment

The population size of some organisms remains fairly constant from year to year, while the populations of other species rise and fall over predictable time periods, usually in response to changes in resources or natural enemies. When food is plentiful, predation is low or disease is absent, populations can swell, but will shrink again when these controlling factors reappear. Typically, organisms with cycling populations are smaller, with fast generation times – larger animals don't reproduce quickly enough to respond to fluctuations in the environment in the same way. Phytoplankton, tiny micro-organisms in the sea, reproduce very quickly when nutrients become available. In the coastal waters of the Bay of Bengal, after the Indian monsoon, the seawater is rich in nutrients from the muddy run-off from the rain. The phytoplankton proliferate and can build up to populations so large they can be seen from space as swirls of colour in the oceans. These high populations don't persist for long and the blooms disappear as the phytoplankton die and sink, but they will go through another population cycle when the next influx of nutrients arrives. Other organisms with cycling populations include small mammals, such as some mice and rabbits, and many invertebrates, including species of butterfly and grasshopper.

RELATED TOPIC
See also
LIMITS TO POPULATION GROWTH
page 52

3-SECOND BIOGRAPHY
CHARLES JOSEPH KREBS
1936–
Canadian ecologist who showed how the natural enemies of snowshoe hares influenced their decade-long population cycles.

30-SECOND TEXT
Julia Cooke

3-SECOND SURVEY
Some populations remain stable, while others peak and plummet – otherwise known as cycling – over regular timescales in response to changes in the environment around them.

3-MINUTE STUDY
The reason for cycling in many populations remains unknown. It is thought that in parts of Europe when vole numbers build up to high densities, the silicon levels in the grasses they eat increases as an induced response to the high grazing load. This may then reduce vole numbers as the grass becomes less digestible. However, more research is needed to be sure it is the silicon defences in grasses that are driving the cycling populations in voles.

The population sizes of species of small organisms on land and in water change dramatically but regularly over time.

METAPOPULATIONS

the 30-second treatment

3-SECOND SURVEY
Metapopulations are spatially isolated populations of the same species that interact infrequently through the migration of individuals between them.

3-MINUTE STUDY
Important in ecology, the role of metapopulations can also be valuable to understand in many industries, such as farming and fishing. For example, if a local population of fish is removed through intensive fishing, it is vital to ensure that sufficient fish remain in the regional metapopulation to replenish stocks. Similarly, for farming practices that rely on pollinators, maintaining healthy metapopulations of pollinating insects across the landscape is important for sustainable crop production.

Metapopulations describe populations of the same species that are separated in space, but close enough for some individuals to move between populations, allowing the populations to interact periodically. At various times different populations may flourish, while others can become extinct, and their interconnectivity can allow them to retain overall stability. In addition, the comparatively small number of individuals that move between groups within the metapopulation allow increased gene flow and can be critical to avoid inbreeding. Populations can be limited by many factors including disease and resource availability. If a population becomes extinct, individuals from an adjacent population can move into the unoccupied space and resume population growth. For example, in savanna ecosystems, annual fires burn parts of the landscape. Depending on the severity of the fire, populations of plants, animals, fungi and bacteria can become locally extinct, and the metapopulations of species are important sources to allow repopulation of a burnt area and to maintain regional species presence. The concept of metapopulations was first described in terrestrial systems, but is highly applicable to aquatic systems, where metapopulations have been observed to be important in maintaining species presence in lakes, reef systems and open water.

RELATED TOPICS
See also
DISPERSAL
page 50

LIMITS TO POPULATION GROWTH
page 52

TROPICAL FORESTS & SAVANNA
page 106

3-SECOND BIOGRAPHY
ILKKA HANSKI
1953–2016
Finnish ecologist renowned for his work showing how metapopulations affect species' population dynamics.

30-SECOND TEXT
Julia Cooke

Infrequent interactions among metapopulations of a species are important for the continued stability of a species overall.

EATING . . .

the 30-second treatment

All organisms need energy to

grow and reproduce. Organisms that can't acquire energy via photosynthesis must eat other organisms. Herbivores most commonly feed on nutrient-rich and easy-to-digest leaves and fruits, but some species have adapted to feed on twigs, nectar, tree bark and even roots. Carnivores have specialized teeth and digestion systems for eating animals. Apex predators such as wolves and sharks are examples of fierce carnivores, but birds, small mammals and even some insects and plants can be carnivorous as well. Different species have evolved different strategies for how and what they eat. Some species are generalists, meaning they are indiscriminate in what they eat. Deer, for example, eat a huge variety of plants, meaning they rarely run out of food options. In contrast, specialists prefer to feed on one or a few prey species. While these species are prone to starvation if their food resource is not abundant, they have the advantage of being adapted to catching and digesting that species when it is around. This means that specialists face little competition from other hungry predators. Eating requires important trade-offs. Time used searching for food is time lost finding a mate or a safe place to sleep. Therefore, species have evolved optimal foraging strategies that allow them to eat just the right amount of food in order to survive.

RELATED TOPICS
See also
. . . AND BEING EATEN
page 62

ENERGY FLOWS
page 74

3-SECOND BIOGRAPHY
CHARLES SUTHERLAND
ELTON
1900–91
English zoologist who outlined key ecological principles associated with food chains and feeding relationships.

30-SECOND TEXT
Stephen Murphy

3-SECOND SURVEY
Organisms that can't produce their own food must eat other organisms in order to survive. Herbivores are species that eat plants; carnivores eat animals.

3-MINUTE STUDY
Different species spend different amounts of time foraging and eating. Some animals with very high energy requirements eat frequently and in large amounts. Hummingbirds, for example, can drink up to half their body weight in nectar in a single day. As such, they spend most of their time finding food. In contrast, large pythons may eat only once every few weeks.

Hummingbirds, with fast metabolic rates, feed frequently on high-energy foods, while large snakes eat infrequently.

. . . AND BEING EATEN

the 30-second treatment

When something is eaten, some of the energy and nutrients from that organism are transferred to the eater. The more nutritious and abundant an organism is, the more sought after it might be, so many living things invest in defences against being eaten. Some plants and animals produce chemicals that taste bad or are poisonous. Others have tough skin, hard shells, spines, prickly hairs or the ability to bite, sting or injure their potential consumers. Animals that are eaten are known as prey, the things that eat them as predators. Predators can also be prey. For example, ants are often ferocious predators, but they are eaten by ant-eating specialists such as echidnas, aardvarks and anteaters. Some organisms try to avoid being detected by the animals that want to eat them, either through camouflage or by mimicking another less palatable or more dangerous species, and many organisms live or shelter where predators can't easily reach them, such as in burrows, hollows and treetops. For some organisms, being eaten, or producing edible parts, is beneficial. Many plants produce tasty fruits around less digestible seeds, so that animals eat the fruit and disperse the seeds in their faeces. Some parasites complete part of their life cycle in animal digestive tracts, and so seek to be eaten.

3-SECOND SURVEY
Being eaten is a threat to many living things, and diverse strategies to avoid or reduce the chance of being eaten have evolved.

3-MINUTE STUDY
Producing defences to avoid being eaten has resulted in some complicated relationships among species. Milkweed plants produce toxic sap that deters most animals from eating them. However, the larvae of the monarch butterfly has evolved to tolerate the toxins and accumulate these in their own bodies to avoid being eaten themselves. The larvae advertise that they are distasteful with bright colours, to deter birds from eating them accidently. This warning of danger is called aposematism.

RELATED TOPICS
See also
DEFENCE & COUNTER DEFENCE
page 36

EATING...
page 60

3-SECOND BIOGRAPHIES
FRITZ MÜLLER
1821–97
German entomologist who showed that noxious species with warning colours tend to look similar, multiplying the effect.

HENRY WALTER BATES
1825–92
British naturalist who first defined ecological mimicry, showing that defenceless species often look like truly dangerous species.

30-SECOND TEXT
Julia Cooke

Ants and milkweed have anti-predator chemical defences, but are still consumed by specialist carnivores and herbivores.

PARASITES

the 30-second treatment

3-SECOND SURVEY
Parasites are organisms that are adapted to live on or in another organism and they can cause disease in their host, or facilitate other diseases.

3-MINUTE STUDY
It is not only animals that have parasites. Parasitic plants have modified roots that penetrate the tissues of host plants to take water and nutrients. Mistletoes are hemiparasites, deriving part of their energy from their hosts but also from photosynthesis. *Rafflesia*, which produce some of the largest flowers in the world, are entirely dependent on their hosts for their energy (holoparasites). Some parasitic plants cause disease symptoms in plants, and expedite infections from other pathogens.

A parasite is an organism that lives in or on another organism, its host. Not all parasites cause diseases and some have very little effect on their hosts. However, some parasites, including species of the single-celled microbe *Plasmodium*, spend part of their life cycle in blood-feeding insect hosts, and other periods in mammalian hosts, where they move when the insect feeds. *Plasmodium* parasites damage the red blood cells of their mammalian hosts, which in humans can cause malaria. In some cases, parasites can affect the behaviour of their hosts to facilitate the next stage of their life cycle. Parasitic fungi in the *Ophiocordyceps* genus release chemicals into their hosts, *Camponotus* worker ants, which cause the ants to move away from their colony and attach to a leaf in an area of high humidity before dying, thereby creating ideal conditions for the fungus to reproduce. There are examples of parasites causing orb-weaving spiders to make a different type of web for them and wood crickets to drown themselves in water. When parasites affect the behaviour of their hosts this dramatically, their hosts are described as zombies. Organisms have evolved defences against parasites. At various times different populations may flourish, while others can become extinct, and their interconnectivity can allow them to retain overall stability.

RELATED TOPICS
See also
DEFENCE & COUNTER
DEFENCE
page 36

VECTORS OF DISEASE
page 126

3-SECOND BIOGRAPHIES
SIR ALGERNON THOMAS
1857–1937
New Zealand biologist who studied the multi-host life cycle of the parasitic sheep liver fluke simultaneously, but independently, to German zoologist Rudolf Leuckart.

LEE HONG SUSAN LIM
1952–2014
Malaysian parasitologist who specialized in flatworms parasitic of fish and described more than 100 new species.

30-SECOND TEXT
Julia Cooke

Parasitic mistletoes have just one host during their life, while microbial parasites move between mosquitoes and people.

COMPETITION

the 30-second treatment

No resource is unlimited, and so organisms that use the same resource are often in competition. Intraspecific competition occurs when a species competes with other individuals of the same species; interspecific competition describes rivalry between species. Competition sees better survival of individuals with traits that increase their relative ability to acquire limited resources, which are then more likely to successfully survive and reproduce, passing on these traits in their genes. For example, animals that can move faster, grow more quickly, survive longer without water, hunt more effectively, or whichever trait gives them an advantage in terms of survival, will then see this trait that affects the outcome of competition passed on. Competition can also result in the formation of new species. Galapagos finches are a classic example, where a small number of finches reached the isolated islands and as the population grew, individuals competed for the seeds they could eat. Gradually, some birds evolved variations in their beak size and shapes, allowing them to specialize in eating different foods. This made them better at competing for resources, resulting in the evolution of more than a dozen species. There is not a single best competitive approach, but rather organisms combine a suite of traits that give them a unique strategy in competitive conditions.

3-SECOND SURVEY
All resources are finite, and so organisms inevitably compete for these among others of the same species – and with other species – to survive and reproduce.

3-MINUTE STUDY
Competition within a species can be fierce. When numerous seeds germinate together, the seedlings compete for space, water, nutrients and light, with only the strongest seedlings surviving. Plants of the same species but different ages also compete. Young trees in dense forests are often outcompeted for light and water by bigger trees. But when a large tree falls and the competition is removed, small trees can use the space and light made available.

RELATED TOPICS
See also
MAKING NEW SPECIES
page 16

DISPERSAL
page 50

3-SECOND BIOGRAPHY
BARBARA ROSEMARY GRANT
& PETER RAYMOND GRANT
1936–
British researchers who have studied the evolution of Darwin's finches since 1973, profoundly changing our understanding of adaptation.

30-SECOND TEXT
Julia Cooke

Different traits can give individuals or species a competitive edge, to tolerate drought, catch prey or find enough food.

MUTUAL BENEFITS

the 30-second treatment

Relationships between organisms often benefit one at the cost of the other, such as in predator–prey interactions, but some relationships have mutual benefits with advantages for both organisms. Mutual benefits typically describe relationships between two different species that have evolved over long periods. Mutualistic relationships can include the provision of food and nutrients; habitat; parasite or disease reduction; defence or shelter from predators; nurseries; pollination; and transport. Clown fish benefit from the protection of the poisonous anemones in which they live, while the fish defend them from predators, eat parasites and supply nutrients. Some acacia species, in Central America, produce hollow thorns as living space for an ant genus, as well as producing packets of protein called Beltian bodies, as food for the ants. In return, the ants actively defend the tree from insect and mammalian herbivores. A challenge for many plants is nitrogen acquisition, and some plants have bacteria in their roots that can fix nitrogen from the air and, in turn, the plant supplies sugars and shelter. Mutualistic relationships are common among distantly related species (such as a plant and animal, or bacteria and animal), and examples of these mutually beneficial relationships are found in all ecosystems.

3-SECOND SURVEY
Not all ecological relationships are antagonistic, as there are many examples of two very different species that have evolved to help each other survive and reproduce.

3-MINUTE STUDY
Most insect-mediated pollination relationships are examples where species derive mutual benefits. Plants are predominantly sedentary, hence a relationship with mobile animals allows effective pollen transfer. Many flowers produce nectar as food for pollinators, while animals move pollen between plants, allowing sexual reproduction. In other systems, animals eat some of the pollen as well as transporting pollen, and in others yet, the flowers provide a refuge for animals to rest or meet to mate.

RELATED TOPIC
See also
POLLINATION & SOCIETY
page 122

3-SECOND BIOGRAPHY
PIERRE-JOSEPH VAN BENEDEN
1809–94
Belgian zoologist who introduced the term mutualism in 1876.

30-SECOND TEXT
Julia Cooke

Clown fish and anemones are a classic example of a mutual relationship where each defends the other from different threats.

COMMUNITIES & LANDSCAPES

COMMUNITIES & LANDSCAPES
GLOSSARY

abiotic Of or having to do with non-living materials or processes. Often used to indicate environmental variables such as soils or light.

atmospheric molecules Molecules making up Earth's atmosphere, including nitrogen and oxygen.

bioavailable Able to be used by living organisms.

biogeography The study of the distribution of species and ecosystems across the Earth.

biomass A measure of the total amount of organic material in a sample, usually measured as dry weight.

bycatch species Non-target species that is inadvertently harvested by commercial fisheries operations.

Calvin cycle Component of photosynthesis through which carbon dioxide is converted into sugar. The sugars produced via the Calvin cycle are used as metabolic fuel.

climax forest Mature forest community with a relatively stable species composition.

colonization The process of an organism moving to a new place, becoming established there and surviving.

disperser An organism that moves from its place of birth. Seeds, spores and young animals often disperse to new habitats.

equilibrium theory An ecological theory predicting that ecological communities return to a steady (i.e. natural) state after a disturbance event.

forest stand A small patch of forest where there are trees that share certain characteristics, such as species, age and size.

heterogeneity Variability across space or time.

hydrothermal vent Deep crack in the ocean floor where geothermal energy is released. Hydrothermal vents can support a variety of species of unique marine life that are able to tolerate the extreme conditions that typify these sites.

macronutrient Essential nutrient required for plant life. Macronutrients are taken up by plants directly from soils in large quantities relative to micronutrients. Examples of important macronutrients include nitrogen, phosphorous and potassium.

mesopredator Predator that occupies intermediate trophic levels. Examples include foxes and small raptors.

micronutrient Essential nutrient required for plant life, but only in a very small quantity. Examples of important micronutrients include calcium and iron.

natural enemy Species that harms another species in some way.

niche Either the set of environmental conditions conducive to a species' survival or the role the species plays in its ecosystem.

nitrogen fixation The process through which atmospheric nitrogen is converted into bioavailable forms such as ammonia and urea. Nitrogen fixation is most commonly performed by specialized bacteria living in soils and in some plant roots.

nutrient pool Source of nutrients within an ecosystem. Nutrients can be stored in soils, water or in living organisms themselves.

photosynthesis A process by which sunlight is converted into usable energy.

senescent Nearing the final stages of life; close to death.

substrate Soil material on which microorganisms can survive and plant roots can take hold.

understory The layer of vegetation in a forest that is found underneath the canopy. The understory layer includes grasses and wildflowers, as well as shrubs and small trees.

ungulate Group of hooved mammals, including deer. Ungulates are important herbivores in many different ecosystems across the globe.

windthrow The process by which large trees are knocked over by high winds. Windthrow events create small canopy gaps where ecological succession occurs – the species structure changes.

ENERGY FLOWS

the 30-second treatment

For life on Earth, the sun is the engine of creation. Indeed, nearly all of the energy that flows through ecological systems originates from sunlight. Organisms capable of directly converting sunlight into metabolic energy via photosynthesis are called autotrophs. These organisms form the base of the food chain and are referred to as primary producers. In most ecological systems, plants are the most important primary producer, although certain types of bacteria, as well as red and brown algae, fill this role as well. In contrast, heterotrophs are organisms that must acquire energy via the consumption of organic matter, usually by eating other plants or animals. Animals that eat primarily plants are called primary consumers, while carnivorous animals are known as secondary consumers. Together, these different organisms form a food chain, or sometimes a web, with energy flowing from producers to consumers. The transfer of energy from one level of the food chain to the next is inherently inefficient, with a large proportion of the total energy being lost as heat. It is for this reason that apex predators such as sharks and wolves, which lie at the very top of the food chain, are rare in comparison to insects and plants. It is also why these species are at a particularly acute risk of extinction.

3-SECOND BIOGRAPHIES
CHARLES SUTHERLAND ELTON
1900–91
British ecologist and zoologist who developed many foundational concepts in ecology, including that of the food chain and invasive species.

MELVIN ELLIS CALVIN
1911–97
American biochemist and Nobel laureate who explained a key part of photosynthesis, now named the Calvin cycle in his honour.

30-SECOND TEXT
Stephen Murphy

Food chains reveal the flow of energy from primary producers like plants to consumers such as small herbivores and predators.

NUTRIENT CYCLES

the 30-second treatment

3-SECOND SURVEY

When organisms die, the nutrients in their decaying tissues are released back into the environment, making them available once again for other organisms to use.

3-MINUTE STUDY

Nutrient cycles are commonly altered by human interactions. For example, humans extract nitrogen from the atmosphere and convert it to bioavailable ammonia. The ammonia derived from this process is then used to fertilize crops, which significantly increases the efficiency of agricultural systems. At the same time, excess nitrogen from fertilizers is often washed away from these systems and deposited into lakes and rivers. The harmful effects of these artificial inputs are known as nutrient pollution, or eutrophication.

In addition to energy, organisms require a number of micro- and macronutrients to survive, grow and reproduce. Important nutrients needed for life on Earth include water, CO_2, nitrogen and phosphorous, among many others. Sources of these important nutrients are varied, and include the weathering of rocks and minerals, as well as the conversion of atmospheric molecules into bioavailable forms. In contrast to energy, which is constantly replenished via the process of photosynthesis, nutrient pools remain more or less stable over time. To avoid the depletion of these pools as organisms grow and reproduce, the nutrients in them must be recycled back into the system. In terrestrial ecosystems, plants primarily intake nutrients through the soil. A notable exception is CO_2, which is taken in from the atmosphere through their leaves. Nutrients are then utilized by plants to perform essential functions such as photosynthesis. These nutrients are then passed from producers to consumers, much like the transfer of energy. When organisms die, the process of decomposition occurs, which is often accelerated by specialized bacteria, insects and fungi. Decomposition releases the nutrients bound up in the dead organic matter, where they are then reincorporated into the soil or transported to other regions. These nutrients then become available again for other plants.

RELATED TOPIC

See also
ENERGY FLOWS
page 74

3-SECOND BIOGRAPHIES

ROBERT H. BURRIS
1914–2010
American biochemist and member of the National Academy of Sciences who revolutionized our understanding of nitrogen fixation.

PETER VITOUSEK
1949–
American ecologist and member of the National Academy of Sciences who studies how humans have altered the nitrogen cycle.

30-SECOND TEXT

Stephen Murphy

Nutrients essential to life on Earth are recycled through a complex interplay between soils, microbes, and plants and animals.

KEYSTONE SPECIES

the 30-second treatment

3-SECOND SURVEY
A keystone species is one that supports and maintains the proper functioning of an entire ecosystem.

3-MINUTE STUDY
Alterations to the population abundances of apex predators can have multiple indirect effects on organisms occupying lower levels of the food chain. Such effects are called trophic cascades, since they reflect a domino-like effect from higher to lower levels. Today, a global decline in the diversity and abundance of top predators has led to dramatic changes to a wide range of ecosystems, a phenomenon known as trophic downgrading.

While sharks, beavers, otters and wolves may not bear any obvious resemblance to one another, they all share a common feature: they are keystone species. Keystone species are those that have a large effect on the ecosystems that they inhabit, despite taking up a relatively low proportion of the total biomass. Importantly, keystone species drive a large number of both direct and indirect ecological interactions with other species, the loss of which can result in significant changes to the ecosystem as a whole. Large predators at the top of the food chain are commonly thought of as keystone species, since they have cascading effects on species that occupy lower levels of the food chain. Small changes in the population abundances of top predators can thus have substantial effects on other species, including primary producers. A well-known example of a keystone species is the wolf, which helps keep deer and elk populations in check. Without wolves, deer and elk increase substantially in abundance, which leads to multiple negative impacts on vegetation and stream ecosystems. In many cases, the removal of keystone species can result in total ecosystem collapse. As such, identifying and protecting keystone species is vitally important for conservation, since saving them can help maintain whole ecosystem function.

RELATED TOPICS
See also
ENERGY FLOWS
page 74

INDIRECT EFFECTS
page 80

3-SECOND BIOGRAPHIES
ROBERT T. PAINE
1933–2016
American ecologist who introduced the influential concepts of keystone species and trophic cascades.

JAMES ALLEN ESTES
1945–
American ecologist who coined the phrase trophic downgrading, where top predators are replaced by mesopredators.

30-SECOND TEXT
Stephen Murphy

By altering the population sizes of their prey, top predators such as wolves have important impacts on ecosystems.

INDIRECT EFFECTS

the 30-second treatment

3-SECOND SURVEY
In nature, the enemy of your enemy can oftentimes be your friend.

3-MINUTE STUDY
Gypsy moth (*Lymantria dispar*) is an invasive species that was introduced to the US in the mid-nineteenth century. Gypsy moths have direct negative impacts on forest trees, since the moth feeds on leaves, causing defoliation of the canopy. In its weakened state, the tree becomes highly susceptible to infections from pathogens, such as the *Armillaria* root rot fungus. In this case, the gypsy moth has an indirect positive effect on the *Armillaria* fungus.

Species compete directly for limiting resources such as food and water, but multiple species across different trophic levels can indirectly mediate these interactions as well. A classic example is the phenomenon of apparent competition, which occurs when multiple species are preyed upon by a common natural enemy. Consider two plant species that are both eaten by the same herbivore. An increase in the abundance of one of the plant species will attract the herbivore to the site, and the presence of the herbivore will then negatively impact both plant species. As a result, the presence of either plant species negatively effects the other, even though the two species do not directly compete. Importantly, humans are common drivers of negative indirect interactions. In commercial fisheries, for example, the presence of an economical fish species can negatively impact non-commercial species that are inadvertently caught. In this way, the presence of the commercial species has an indirect negative effect on bycatch species. Identifying indirect effects in both natural and human-modified systems can be highly challenging given the large number of potentially interacting species. However, understanding indirect effects is extremely important for determining how ecological communities function.

RELATED TOPICS
See also
COMPETITION
page 66

KEYSTONE SPECIES
page 78

3-SECOND BIOGRAPHIES
ROBERT DAN HOLT
fl. 1977–
American ecologist who introduced the concept of apparent competition.

DEBORAH GOLDBERG
fl. 1980–
American plant ecologist who studies direct and indirect interactions in plant communities and their effects on diversity and evolution.

30-SECOND TEXT
Stephen Murphy

A sea turtle is inadvertently caught by a fishing trawler, an example of a negative indirect interaction.

DISTURBANCE & RESILIENCE

the 30-second treatment

3-SECOND SURVEY

Events such as storms can significantly change local environments and, therefore, the species found there; resilience is the community's ability to return to its pre-disturbance state.

3-MINUTE STUDY

Disturbances are not exclusively abiotic in nature. Biological disturbances caused by pest or pathogen outbreaks, as well as by invasive species, are also common. Some forests are regulated by intermittent insect outbreaks that alter understory light and reduce the growth rates of abundant species. However, because natural communities are not adapted to invasive insect outbreaks, these disturbances can cause long-term alterations to communities.

A disturbance is an event that alters key ecosystem properties related to soils, vegetation and resources. Examples of disturbances include wildfires, hurricanes and drought. Disturbances are important because they alter diversity and species composition. While disturbances can be highly destructive, they are also critical for the long-term maintenance of ecological diversity. By creating new environmental conditions, disturbances allow for a wider range of species to coexist within a community. In forests, wind disturbance is especially common, which leads to the toppling of large senescent trees. Such small-scale disturbance events create novel ecological conditions that encourage the recruitment of more light-demanding species. As a result, a forest landscape can be viewed as a patchwork of small stands in various stages of growth. Without these small windthrow disturbances, shade-tolerant species persist indefinitely, resulting in the complete competitive exclusion of other species. Importantly, the ability of a community to adapt to disturbance depends on how frequently the disturbance occurs over time. Disturbances that are outside the range of conditions normally experienced by the community may be particularly destructive. This is why ecosystems are oftentimes not resilient to disturbances caused by humans.

RELATED TOPIC

See also
ECOLOGICAL SUCCESSION
page 84

3-SECOND BIOGRAPHIES

ALEXANDER WATT
1892–1985
Scottish botanist who proposed that small canopy disturbances help to maintain forest diversity and productivity.

STEWARD T. A. PICKETT
1950–
American senior scientist who developed a theory for natural disturbances' effects on ecological dynamic.

MONICA G. TURNER
1958–
American ecologist who popularized landscape ecology and showed that wildfires are essential for the health of many forests.

30-SECOND TEXT

Stephen Murphy

A tree seedling emerges after a devastating storm, highlighting the resiliency of communities to disturbance events.

ECOLOGICAL SUCCESSION

the 30-second treatment

3-SECOND SURVEY
A successional community is one undergoing predictable changes in composition and diversity as a result of recovery from a disturbance event.

3-MINUTE STUDY
The eruption of Mount St Helens, in the United States in 1980, offers a striking example of ecological succession. The explosion effectively destroyed all life within a 160-km (100-mile) radius of the eruption. In some areas, soils were completed removed, leaving only bare rock. Over time, a new soil substrate developed as early colonizers decomposed, allowing the establishment of hardy trees and shrubs. Almost 40 years after the explosion, large fir trees can now be found growing on the mountain.

Plant communities are constantly changing over time. When these changes are predictable, we call this process ecological succession. Succession begins following a disturbance event that destroys some or all of the existing vegetation in a community. Examples include fires, hurricanes or even volcanic eruptions. Such disturbances create environmental conditions that are strikingly different from those found before the disturbance event occurred. Disturbances increase light availability, revitalize soil nutrients, create new microhabitats and remove competitors. These novel environmental conditions promote the colonization, germination and growth of species that would have been unable to persist in the pre-disturbance community. Generally, fast-growing and light-demanding grasses and wildflowers, called pioneer species, dominate early stages of succession. Eventually, these species are overtaken by slower-growing woody shrubs and trees that cast shade on the pioneer species. As succession proceeds, the diversity of the community generally increases as more species become established. Eventually, the slowest-growing and most competitive species become dominant, and the system is maintained until another disturbance event occurs.

RELATED TOPIC
See also
DISTURBANCE & RESILIENCE
page 82

3-SECOND BIOGRAPHIES
HENRY CHANDLER COWLES
1869–1939
Early pioneering ecologist who developed one of the first theories of plant succession through his work on the sand dunes of Lake Michigan.

HENRY ALLAN GLEASON
1882–1975
American botanist who proposed that ecological succession is driven by the responses of individual organisms – controversial at the time, but now widely accepted.

E. LUCY BRAUN
1889–1971
American botanist who described and mapped the climax forest communities of the eastern United States.

30-SECOND TEXT
Stephen Murphy

Wildfires and other disturbances reset plant communities and initiate recovery.

SPECIES–AREA RELATIONSHIP
the 30-second treatment

To say that a larger area contains more individuals would seem as obvious as saying a larger bucket holds more water, but should larger areas also contain more species? In the 1920s, Swedish scientist Olof Arrhenius established a theory for the species–area relationship by developing a prediction based on species' abundances. Decades of work has demonstrated that he was fundamentally correct, and the number of species increases continuously as areas get larger. There are many ideas for why this may be so. It could be that the number of species present grows with area, simply because larger areas hold more types of habitats; that larger areas are bigger targets for dispersers; that larger areas harbour larger populations and so are more resilient to extinction; and that extremely large areas have distinct evolutionary histories and so when joined into a single area contain many more species. A fascinating universal rule arises in cases when smaller areas are completely nested within larger areas (as opposed to being distinct sites, such as islands): the rate at which species number increases with area is a function of the ratio of the number of individuals (animals, plants, fungi or bacteria) and species in the area. This species–area relationship can then be used to predict the loss of species due to habitat fragmentation.

3-SECOND SURVEY
The accumulation of species with increasing area stops only when the entire Earth is considered.

3-MINUTE STUDY
Larger areas don't always have more species. The number of species in an area is also affected by where the place is on Earth. Arctic and temperate regions have fewer species than tropical regions, and the deep ocean has fewer species than coastal waters. For example, the tropical country of Colombia is home to about 1,900 bird species, whereas South Africa, which is roughly the same size but with a Mediterranean climate, has only 850.

RELATED TOPICS
See also
ISLAND BIOGEOGRAPHY
page 88

GLOBAL PATTERNS OF BIODIVERSITY
page 96

3-SECOND BIOGRAPHIES
FRANK W. PRESTON
1896–1989
British–American engineer and ecologist who refined theoretical predictions of the shape of the species–area relationship.

JOHN HARTE
1939–
American conservationist and ecologist who discovered a universal scaling law for species–area relationships.

DANIEL SIMBERLOFF
1942–
American ecologist who demonstrated that remote islands and smaller islands harbour fewer species.

30-SECOND TEXT
Adam B. Smith

The number of species increases as the size of the area increases.

ISLAND BIOGEOGRAPHY

the 30-second treatment

3-SECOND SURVEY
The diversity of species on an island is a balance between colonization, extinction and, in the longer term, speciation, which are influenced by island area, age and isolation.

3-MINUTE STUDY
Island endemics sometimes evolve to be much smaller or larger than their continental counterparts. For example, horse-sized pygmy mammoths on California's Channel Islands survived there for tens of thousands of years before perishing at the same time as their mainland relatives. In contrast, Polynesians settling in New Zealand found moa birds, some of which reached up to 3.6m (12ft) tall, which were predated upon by the world's largest eagle. Both were driven to extinction.

Fumigating entire mangrove islands to kill all insects, and then sawing portions off islands to reduce their size, seems an odd start to a theory that is now fundamental for conserving threatened species, but that is how ideas of island biogeography were first experimentally tested. These experiments demonstrated the fundamental tenets of the theory: the number of species on an island is a balance between rates of species colonization and extinction. These in turn are functions of the remoteness (how likely they are to be colonized) and area of the island (which affects habitat availability and size of populations on the islands). Those islands near sources of species will be found more quickly. If a species arriving on a new island can survive there, it is because it can take advantage of the unoccupied ways of life (niches). As these niches fill, fewer new species can squeeze in, but bigger islands tend to have more niches. As a result, large islands (or patches of habitat) near sources of potential colonizers should have more species than remote, small islands. Of course, things are a little more complicated. Remote islands are often evolutionary hotspots, as the lucky species finding the island rapidly diversify as they fill vacant niches (for example the Galapagos finches).

RELATED TOPICS
See also
HABITATS & NICHES
page 22

SPECIES–AREA RELATIONSHIP
page 86

3-SECOND BIOGRAPHIES
EDWARD O. WILSON
1929–
American biologist who, along with R. H. MacArthur, formulated the theory of island biogeography.

ROBERT H. MACARTHUR
1930–72
Canadian–American ecologist who formulated the mathematical and empirical basis of the equilibrium theory of island biogeography.

30-SECOND TEXT
Adam B. Smith

Fumigation of islands and examination of fossil evidence have been critical to understanding how colonization and extinction affect island biodiversity.

7 April 1930
Born in Ontario, Canada

1953
Receives master's degree
in Mathematics from
Brown University

1957
Earns PhD from
Yale University; his
dissertation on warblers
becomes a classic study
in ecology

1958
Takes his first faculty
position at the University
of Pennsylvania

1965
Moves to Princeton
University and remains
there for the rest of
his career and life

1967
Publishes the theory
of island biogeography
with Edward O. Wilson
of Harvard University

1969
Elected to the National
Academy of Sciences

1972
Publishes his final
work, *Geographical
Ecology: Patterns in the
Distribution of Species*

1 November 1972
Dies at the early age
of 42 from cancer

ROBERT H. MACARTHUR

Perhaps more than any other individual, Robert Helmer MacArthur shaped the way modern ecology is conducted today. His lasting contributions to the field stem from his then revolutionary strategy of merging ecological theory with mathematical models and observational field data. He used this framework to reshape the way ecologists understand many foundational concepts, including biodiversity, species distributions and ecological interactions.

MacArthur was born into an academic family in Toronto. He spent his undergraduate and early graduate career studying mathematics before moving to Yale to earn his PhD in Biology. His dissertation focused on how closely related bird species partition environmental resources through behavioural differences. He also introduced a general theory for the origin and maintenance of species abundance distributions. The research published from his time at Yale was widely read at the time, earning him a position on the Biology faculty at the University of Pennsylvania in 1958.

Famous for going against the established norms of his day, MacArthur believed that all empirical studies should be driven by clearly defined hypotheses based on ecological theory. He also did not follow the standard protocols for scholarship. Specifically, he was not afraid to present bold new ideas that he knew would be open to criticism. And while this strategy earned him plenty of detractors, the legacy of his approach lives on to this day.

In 1965, MacArthur was invited to join the faculty at Princeton University. Soon after, he teamed up with E.O. Wilson of Harvard for what would become one of the most important collaborations in the history of ecology. Together, they developed the theory of island biogeography, which explained variation in the number of bird species across oceanic islands. The theory offered a general framework for understanding species assemblages and patterns of diversity that continues to have a significant impact on the fields of ecology and conservation biology today.

The reach and impact of MacArthur's work cannot be understated, and his ideas are still debated and written about. Unfortunately, he was diagnosed with renal cancer in 1971, and died the following year at the young age of 42. The knowledge of his impending death inspired him to synthesize his views on the patterns and distributions of biodiversity in his final work, *Geographical Ecology*.

Stephen Murphy

BIOMES & BIODIVERSITY

BIOMES & BIODIVERSITY
GLOSSARY

benthic An ecological zone in freshwater and marine environments at the bottom of a body of water, including the sediment and sub-surface layers.

biome A major region that is defined by the plants and animals living there and determined largely by climate and latitude. A biome differs from an ecosystem, which encompasses organisms and their interactions with each other and the environment, in a specific location. For example, tropical forest is a biome that includes ecosystems such as montane humid forest, seasonal dry forest and coastal-plain swamp forest.

ecosystem services These are the direct or indirect benefits to people provided by ecosystems. These services may impact human survival or quality of life and are categorized into four service types: provisioning, regulating, supporting and cultural.

elevational diversity gradient An observed pattern in ecology where an increase in elevation leads to an increase in species richness, up to a certain point, where it begins to decrease again. This leads to a hump-shaped trend in diversity in mountain ranges, where the highest numbers of different species are found at mid-elevations.

endemic/endemism Relating to an organism that is found only in a specific location.

glaciation The process by which glaciers are formed as well as the movement and recession of glaciers and how they shape the surrounding landscape. In the past, glaciers covered a much larger extent of the Earth's surface. A small glacier, less than 0.1 km² in size, is known as a glacieret.

latitudinal diversity gradient
A biogeographic pattern that describes the high diversity of species at the equator with declining biodiversity towards the poles.

lentic A body of standing freshwater that includes a variety of systems such as ditches, ponds, seasonal pools, marshes, wetlands and lakes.

lotic Any form of continuously flowing water body from small springs and streams up to large rivers.

ocean acidification The decrease in ocean pH caused by increasing levels of atmospheric carbon dioxide (CO_2) due to human activities such as burning fossil fuels.

palaeontologist A scientist who uses fossils to study the history of life on Earth.

pelagic An ecological zone referring to open water away from the coast or ocean floor.

photosynthesis A process performed by plants, algae and other microorganisms that uses energy from sunlight, combined with water and carbon dioxide, to create chemical energy in the form of glucose.

prokaryote Microscopic single-celled organism of the biological domains Archaea and Bacteria. Such organisms lack a distinct nucleus (meaning DNA is not bound within a membrane) or other membrane-bound compartments with specialized functions called organelles.

trophic levels Categories used to define an organism's position in the food chain. These are broadly classified as producers, consumers and decomposers.

tundra The tundra biome is characterized by a lack of trees and is found in the Arctic and at high elevations in mountains.

GLOBAL PATTERNS OF BIODIVERSITY

the 30-second treatment

Biological diversity, shortened

to biodiversity, describes the variety of living things inhabiting the Earth, from microscopic bacteria to giant redwood trees. Biodiversity exists at many levels from genes, to species, and all the way up to ecosystems. Biodiversity is not uniformly distributed across the globe. Most terrestrial and aquatic organisms follow a latitudinal diversity gradient – the pattern that biodiversity is lowest at the poles and increases towards the equator. On land, diversity is highest for most organisms at middle elevations – an elevational diversity gradient. In the ocean, species richness decreases with depth. Latitude, elevation and depth are the main drivers of biodiversity patterns. These factors are interconnected with the energy available to organisms in the form of temperature and precipitation. Current patterns of biodiversity may also reflect the evolutionary history of mass extinctions and subsequent rapid diversification events. For example, when dinosaurs became extinct there was a higher diversification of mammal species because most competitors and predators had been removed. In fact, it seems that only 200,000 years after the mass extinction of dinosaurs, the first ancestors of mammals switched from being nocturnal to being active in the daytime, which helped them become as diverse as they are today.

3-SECOND SURVEY
An estimated 5 to 20 million species make up the world's biodiversity and are distributed globally according to latitude, elevation and depth.

3-MINUTE STUDY
Biodiversity hotspots are regions of the world that contain high diversity (numbers of species) and high levels of endemism (unique species found in only one area), combined with being extremely threatened. There are 36 hotspots that cover 2.4 per cent of the Earth's surface. Yet organisms within this tiny area provide 35 per cent of the world's ecosystem services, containing 43 per cent of endemic bird, mammal, reptile and amphibian species as well as over half of plant endemics.

RELATED TOPICS
See also
VALUING BIODIVERSITY
page 116

ECOSYSTEM SERVICES
page 118

THE SIXTH MASS EXTINCTION
page 138

3-SECOND BIOGRAPHIES
ALEXANDER VON HUMBOLDT
1769–1859
German explorer whose work was the foundation for the field of biogeography.

E. O. WILSON
1929–
American ant biologist who first used the term 'biodiversity' in a scientific book.

30-SECOND TEXT
Heather Campbell

Biodiversity hotspots have high numbers of endemic species, such as proteas in the Cape floristic region and the lemurs of Madagascar.

OCEANS

the 30-second treatment

The World Ocean combines the Pacific, Atlantic, Indian, Arctic and Southern Oceans to cover 70 per cent of the Earth's surface. Life in the ocean is governed by three interconnected gradients in solar energy, water depth and the change in environment from coastal to open water. Light can penetrate the water from a few centimetres to 250 metres (820 feet), yet life persists even in the darkest 11,000-metre (36,000-feet) trench. Marine organisms inhabit open oceans (pelagic) and the bottom sediment and rock (benthic). Other organisms are specially adapted to live at the air-water interface, like sea skaters, the only open-ocean insects, or the carnivorous and highly venomous Portuguese man-of-war. The top layer of ocean water is crammed with phytoplankton. These are photosynthetic creatures that obtain their energy from the sun, producing half of the world's oxygen and forming the basis of the marine food chain. 'Plankton' describes any organism that drifts on currents such as bacteria, archaea, algae and protozoa, and includes the tiniest ocean life form – a marine virus measuring less than a micrometre (smaller than 1 millionth of a metre). At the opposite end of the scale, oceans are home to the largest animal on the planet, the blue whale, which can measure 30 metres (100 feet) and weigh up to 150 tonnes.

3-SECOND SURVEY
Although less than 5 per cent of the oceans have been explored, they contain a third of known species.

3-MINUTE STUDY
Corals are marine invertebrates that live in colonies made up of individual animals called 'polyps'. Coral reefs are biodiversity hotspots, containing as many as 1,000 species per square metre (10 square feet). They are comparable in richness of species to tropical forests. One fifth of coral reefs have been destroyed and 60 per cent are at risk due to destructive human activities such as mining, pollution, fishing and ocean acidification, and climate change.

RELATED TOPICS
See also
FRESHWATERS
page 100

VALUING BIODIVERSITY
page 116

3-SECOND BIOGRAPHIES
FREDERICK WILLIAM HERSCHEL
1738–1822
German-British astronomer who showed that coral is an animal, not a plant.

JOHN MURRAY
1841–1914
Scottish-Canadian oceanographer who discovered the mid-Atlantic ridge and oceanic trenches.

YULY MIKHAILOVICH SHOKALSKY
1856–1940
Russian oceanographer who coined the term 'World Ocean'.

30-SECOND TEXT
Heather Campbell

Marine organisms exhibit many different sizes and forms – from delicate corals to enormous whales.

FRESHWATERS

the 30-second treatment

3-SECOND SURVEY
Streams, rivers, floodplains and lakes all make up the freshwater biome, referred to as lentic (still water) and lotic (running water).

3-MINUTE STUDY
Lake Baikal in Russia claims a host of freshwater lake world-record titles. At 1,637 metres (5,371 feet), it is the deepest lake, and by volume it is the largest at 23,600 cubic kilometres (5,670 cubic miles) – holding 20 per cent of the world's fresh surface water. It is also the most ancient, estimated to be 25 million years old. It is home to the nerpa, the world's only freshwater seal, as well as 2,000 other plants and animals, of which two-thirds are unique to Lake Baikal.

Lakes and rivers contain a tiny

0.01 per cent of all water on Earth. There is less biodiversity in freshwaters than in the ocean or on land – around 10 per cent of described species – but if the smaller available area of freshwater is taken into account, then these habitats are actually one of the most species-dense biomes. Water levels in lakes and rivers are determined by climate but fluctuate rapidly with changing weather. These constant changes require freshwater communities to be highly tolerant of varying habitat conditions. Continuous disturbance leads to rapid speciation, as seen in the high diversity of freshwater fish. Freshwater organisms evolved from ocean-dwellers via two pathways. For insects, plants, birds and mammals, colonization followed evolution on land, with secondary colonization of freshwater from land ancestors. Prokaryotes, algae, crustaceans, most molluscs and fish, however, colonized freshwaters directly from the ocean. The larger the water body, the more freshwater species it holds. This is due to greater structural complexity and a variety of niches, as well as stability in the face of changing geology and climate. Freshwater biodiversity increases towards the equator because of a longer growing season, abundance of food and high turnover of generations near the tropics.

RELATED TOPICS
See also
GLOBAL PATTERNS OF BIODIVERSITY
page 96

OCEANS
page 98

3-SECOND BIOGRAPHIES
PIERRE PERRAULT
1608–80
French amateur scientist who developed the concept of the water cycle.

(ROBERT) ANGUS SMITH
1817–84
Scottish chemist who discovered that air pollution caused acid rain.

FRANK MORTON CARPENTER
1902–94
American palaeontologist who described the largest insect to have ever lived, the extinct dragonfly *Meganeuropsis permiana*.

30-SECOND TEXT
Heather Campbell

Freshwater species invest less energy in interactions than ocean species, meaning fewer bright breeding colours.

TUNDRA & BOREAL FORESTS

the 30-second treatment

3-SECOND SURVEY
Tundra and boreal forests are northern or high-elevation biomes defined by extreme cold and long winters that limit vegetation growth, resulting in low biodiversity.

3-MINUTE STUDY
Tundra means 'treeless tract'. Tree growth is prohibited by low temperatures, short summers and permafrost. Permafrost is permanently frozen ground preventing deep root systems. Many Ice Age animals have been discovered preserved in permafrost, including woolly mammoths, horses, lions, bison and woolly rhinos. Plant tissue buried 30,000 years ago by an Ice Age squirrel was revived by scientists in the laboratory to grow new plants.

South of the permanent polar ice, at latitudes of 55 to 70 degrees North, is Arctic tundra. Head south again and you'll find the tree line, an ecological boundary zone between the tundra and boreal forest biomes. Boreal forests, also called 'taiga', are dominated by conifers such as pine, spruce, larch and fir, as well as some deciduous birch and poplar, which reach the highest latitudes of any trees on Earth. Tundra vegetation consists of small shrubs, grasses, mosses and lichens, with similar plants found in high-elevation alpine tundra and Arctic tundra. Tundra was formed 10,000 years ago, making it the world's youngest biome. Tundra and boreal forests share some species between North America and Eurasia but collectively have low biological diversity that decreases towards the North Pole. These biomes experience long, cold, dark winters, so organisms living here are adapted to low temperatures and short growing seasons during the brief summer months. Many animals survive the winter by hibernating, for example, marmots and Arctic ground squirrels, while others, such as caribou, migrate to warmer areas. Top mammalian predators will travel large distances to hunt their migratory prey, with grey wolves, black bears and Siberian tigers able to cover thousands of miles in one year.

RELATED TOPICS
See also
TEMPERATE FORESTS & GRASSLANDS
page 104

TROPICAL FORESTS & SAVANNA
page 106

3-SECOND BIOGRAPHY
GEORGES CUVIER
1769–1832
French palaeontologist who recognized woolly mammoth remains as an extinct species rather than a modern elephant.

30-SECOND TEXT
Heather Campbell

The vast and bleak northern biomes are inhabited by only the hardiest of cold – and dark – adapted organisms.

TEMPERATE FORESTS & GRASSLANDS

the 30-second treatment

3-SECOND SURVEY
Accounting for a fifth of terrestrial land surface area, temperate biomes consist of broadleaf and mixed forests, coniferous forests and grasslands.

3-MINUTE STUDY
Grasslands are a prime example of trophic levels in an ecological pyramid. As producers, plants appear at the bottom of the pyramid. Herbivores feed on the plants and are primary consumers. These are then eaten by secondary consumers – the carnivores. This system remains balanced because at each level of the pyramid not everything gets eaten. Herbivores eat around 10 per cent of the plant material and predators eat 10 per cent of the herbivores.

Temperate forests and grasslands are found in both hemispheres, in areas with warm, dry summers and cool, wet winters. These biomes have moderate levels of biodiversity and are the most well-studied biomes in ecology. Forests are dominated by trees forming a more or less continuous canopy. Human activities have had a significant impact on the nature and extent of temperate forests with primary (old-growth) forest cover varying from less than 1 per cent in Western Europe and North America to around 25 per cent in New Zealand. Secondary forests grow after land is reclaimed following logging or agriculture. If the climate is too dry for forest to grow then there will be temperate grassland consisting of grasses and herbs. Around the world, temperate grasslands are known as prairie in North America, steppe across Eastern Europe and Asia, pampas in South America, and veld in South Africa. In Australia there are shrublands and temperate savanna. Growth of grassland plants is stimulated by fire and herbivore activity. Grasses produce leaves from the base of the plant, so when a grazing animal eats the upper part of the leaf, a new leaf grows in its place. Trampling by large herds of grassland animals such as bison and antelope can also stimulate plant growth.

RELATED TOPIC
See also
TUNDRA & BOREAL FORESTS
page 102

3-SECOND BIOGRAPHIES
ARTHUR TANSLEY
1871–1955
English botanist who founded the world's first ecological society to study the vegetation of the British Isles.

CHARLES ELTON
1900–91
English population biologist who outlined the idea of ecological pyramids and trophic levels.

MELVIN CALVIN
1911–97
American biochemist awarded the Nobel Prize in Chemistry for discovering photosynthesis.

30-SECOND TEXT
Heather Campbell

Apex predators, such as the grey wolf, are secondary consumers that occupy the highest trophic level of an ecological pyramid.

TROPICAL FORESTS & SAVANNA

the 30-second treatment

3-SECOND SURVEY
Tropical forests and savannas are the most diverse and unique biomes on the planet, containing half of the 5 to 20 million estimated species.

3-MINUTE STUDY
The Janzen–Connell hypothesis explains high tree diversity in tropical rainforests. Near a parent (seed-producing) tree there are a lot of seed-eating herbivores, meaning that most seeds will be found and eaten, or destroyed by diseases and pests carried by seed-eating animals. A seed further away from the parent tree stands a better chance of surviving and growing into a sapling. This process prevents parent and offspring trees growing next to each other, so no species becomes common in an area.

In the tropics, vegetation

structure changes along a rainfall gradient, from rainforest in the wettest areas, to monsoon and dry forest where precipitation is lower and more seasonal. Forest type also varies with latitude, elevation and soil type. Tropical forests can have as many as 1,000 tree species in one square kilometre. They are characterized by high species diversity, a high frequency of cross-pollination, widespread mutualisms and high rates of energy flow. Tropical forests support 350 million people, including 60 million indigenous people, who use forests intensively for subsistence and survival. Forests provide a vast range of ecosystem services such as climate regulation, pollution control, soil protection, nutrient cycling, water regulation and carbon storage. Deforestation is particularly problematic in tropical forests, including the largest continuous remaining tropical forests in the Amazon and Congo Basin. For areas without enough rainfall to support extensive tree cover there is wet, dry and thorn-bush savanna. Savanna describes a grassland with sparsely scattered trees, which act as keystone species and include the iconic African thorny acacias. Savannas support high densities of herbivores, including one of the largest mammal gatherings on Earth – the 1.3 million individual wildebeest that migrate across the Serengeti.

RELATED TOPICS
See also
ECOSYSTEM SERVICES
page 118

CLIMATE CHANGE
page 140

3-SECOND BIOGRAPHIES
MARIA MERIAN
1647–1717
German scientific illustrator who documented tropical insect life cycles.

YNÉS MEXÍA
1870–1938
American–Mexican botanist who collected around 150,000 tropical plant specimens, including 500 new species.

JOSEPH CONNELL
1923–
American ecologist who studied high species biodiversity in Australian rainforest and coral reefs.

30-SECOND TEXT
Heather Campbell

Many of the world's last remaining mega-herbivores, including giraffes and elephants, live in savanna.

1 April 1940
Born in Nyeri, Kenya

1960–66
Studies in the US for a
BSc and MSc in Biology

1967–69
Studies at the University
of Giessen and
the University of Munich

1971
Awarded a PhD in
Veterinary Anatomy from
the University of Nairobi

1976
Appointed Chair of the
Department of Veterinary
Anatomy at the
University of Nairobi

1977–2011
Founder and Chair of the
Green Belt Movement

2002–2007
Serves as Member of
Parliament and Assistant
Minister for Environment
and Natural Resources in
Kenya

2004
Awarded the Nobel Peace
Prize

2005
Appointed Goodwill
Ambassador to the Congo
Basin Forest Ecosystem

2006
Spearheads the United
Nations Billion Tree
Campaign

2010
Founds the Wangari
Maathai Institute for
Peace and Environmental
Studies

25 September 2011
Dies in Nairobi, Kenya

WANGARI MAATHAI

Wangari Muta Maathai – the

first African woman to win a Nobel Prize – was a Kenyan environmentalist most famous for founding the Green Belt Movement. Growing up in Kenya, Maathai excelled at school, which won her a scholarship to study for a Biology degree in 1960 at Mount St Scholastica College, Kansas. Following her undergraduate studies she stayed in the US to obtain her master's degree at the University of Pittsburgh in 1966. After studying for her doctorate in Anatomy at two German universities she returned to Kenya, basing herself at the University of Nairobi. Here, she went on to claim a remarkable number of firsts: first East African woman to receive a PhD; then first female chair; followed by first female associate professor.

While working in veterinary anatomy at the University of Nairobi, Maathai became involved in environmental causes, particularly recognizing the role of environmental restoration and job creation. She began by getting women to plant trees, and this activism eventually led to her establishing the Green Belt Movement. Today, the movement is responsible for the planting of over 51 million trees by a workforce of 30,000 women trained in forestry, food processing and bee-keeping. The project not only generates an income in rural areas of Kenya, but also helps the environment by combating deforestation and soil erosion.

Maathai served in the Kenyan Parliament but her greatest achievement remains her award of the Nobel Peace Prize in 2004 for her 'contribution to sustainable development, democracy and peace'. She worked on many environmental projects and was the inspiration for a United Nations campaign to combat global warming and biodiversity loss by planting a billion trees. This number was exceeded in the first year and now stands at a grand total of 13.6 billion trees planted globally – it was so successful that the project has been renamed the Trillion Tree Campaign. In 2006, the same year as launching the Billion Tree Campaign, she met with Barack Obama and planted a tree in Uhuru Park in Nairobi.

Despite having no formal training as an ecologist or forester, Maathai emphasized the links between deforestation and environmental degradation. Her recognition of the link between human well-being, economics and the environment, as well as the role of forests in maintaining a healthy planet, show that her impact is beyond that of a politician or activist. In 2011, Maathai died of complications from ovarian cancer. Her legacy is felt beyond the African continent on which she was born. Speaking about her Nobel Peace Prize award, she said that the committee were 'sending a message that protecting and restoring the environment contributes to peace; it is peace work.'

Heather Campbell

DESERTS

the 30-second treatment

Deserts are one of the most extreme environments on Earth, but still provide a home to 17 per cent of the human population. Deserts can be classified by rainfall and locality into hot, cold, semi-arid or coastal deserts, with only 10 to 20 per cent covered in sand. They are teeming with life and have the third-highest levels of biodiversity and endemism for any biome. Many organisms have adapted to life in these harsh conditions and extreme temperatures. The *Cataglyphis* ant lives in the Sahara and is one of the most thermo-tolerant land animals, foraging at a body temperature over 50°C (122°F). Many other desert animals keep cool by being active at night and living in burrows, where temperatures are lower. They are often pale in colour to prevent them from absorbing heat from the sun. The large ears of the fennec fox of North Africa help it to dissipate heat. Desert organisms are also adept at collecting and storing water. Plants such as cacti have large water storage capacity and spines so they don't lose water through evaporation. The Australian thorny devil has scales on its body to collect dew that is then channelled down to the corners of its mouth, while a similar strategy is observed in Namibian fogstand beetles, which drink water droplets that condense on their body when standing in fog.

RELATED TOPIC
See also
NATURAL SELECTION
page 14

3-SECOND SURVEY
Deserts cover a fifth of the world's land surface and receive less than 250 mm (10 inches) of rain a year. The largest deserts are the polar Arctic and Antarctic deserts.

3-MINUTE STUDY
Deserts might be the most record-breaking places on the planet. The highest recorded temperature on Earth (56.7°C/134°F) was in Death Valley in the Mojave Desert, while the coldest (–89.2°C/–128.6°F) was at the Vostok Station in the Antarctic Desert. At 200 million years old, Chile's Atacama Desert is a double-winner for title of the oldest desert and the driest non-polar place on the planet. The McMurdo Dry Valleys, Antarctica, claim the record of the driest place on the planet.

3-SECOND BIOGRAPHIES
CONSTANTIN GLOGER
1803–63
German ornithologist who proposed that birds and mammals would be darker in humid and warm environments and, therefore, desert-inhabiting animals tend to be light-coloured.

EMAN GHONEIM
fl. 2000
Egyptian geomorphologist who discovered an 11,000-year-old mega lake (30,750 km²/11,872 square miles) buried beneath the sand in the Sahara Desert.

30-SECOND TEXT
Heather Campbell

Deserts may superficially appear devoid of life but are occupied by a huge diversity of plants and animals.

APPLIED ECOLOGY

biocide A poisonous artificial substance that can kill a wide range of life rather than being limited in range, such as a herbicide or insecticide. Biocides are often used in a non-neutral manner.

bioprospecting The act of searching for biological products with economic value, often for the pharmaceutical industry. Such activities have been criticized by some nations as bio-piracy, because the countries of origin see no benefits from the discoveries resulting from bioprospecting.

biosphere The regions of the planet inhabited by living organisms, and the sum total of all ecosystems.

carbon sequestration The capturing and containment of carbon from the atmosphere. This can be natural (through trees) or artificial, using chemical means to trap carbon dioxide.

carrying capacity The maximum population of a species that a habitat can contain indefinitely. Below this level, populations tend to increase; above it, they decline.

chance demographic effect As a consequence of small population size, random events (such as only males being born one year) can have large effects on a population.

Dichlorodiphenyltrichloroethane (DDT) A synthetic insecticide and perhaps the first truly successful global means of controlling mosquitoes and hence malaria. However, DDT is persistent in the environment, and can accumulate in the tissues of species at higher trophic levels, causing mortality in non-target species.

forage crop A crop grown as fodder for animals, rather than for consumption by people.

genetic drift Random changes in genotypes resulting from chance losses of alleles resulting from small population size. This can lead to the loss of genetic variation in a population and result in the reduced ability of a population to adapt to environmental change.

green economy An economy that aims for low-carbon, sustainable resource use to minimize environmental risk. Most users of the term include elements of social justice, and active intervention in markets, to avoid a focus on short-term profit making over longer-term sustainability.

monoculture Single species (usually a crop) growing in a given area.

neonicotinoid insecticide Synthetic systemic neuroactive chemicals structurally related to nicotine, designed to kill insect pests. These have proved controversial, and there is evidence to suggest that these insecticides play a significant part in pollinator declines.

nutrient cycling The repeated movement of a nutrient from the environment, passing through living organisms and returning back to the environment.

pathogen A microorganism that can cause disease. Pathogens can have massive effects on population sizes, such as the plague, which killed tens of millions of European and Asian people, peaking in western Europe in the mid-fourteenth century.

pesticide A synthetic or naturally occurring chemical used to kill pests. Natural pesticides have been used for centuries, but the rapid growth in artificial pesticides since the Second World War has had enormous environmental consequences.

symbiotic relationship A close interaction between two different organisms that bring mutual benefits. In the closest symbiotic relationships, neither can survive without the other.

urban heat island effect The consequence of human activities causing urban areas to become significantly warmer than surrounding regions. This can be caused by the use of building materials such as concrete, which traps heat and slowly reradiates it, and by the direct release of heat from human activities. Cities may be several degrees warmer than the surrounding region, and other weather patterns, such as rainfall, are also different.

VALUING BIODIVERSITY

the 30-second treatment

In the past 50 years, vertebrate numbers have declined by over 60 per cent, as the biosphere is placed at risk of catastrophic damage. While many readers will agree that biodiversity should be protected for its intrinsic value, few of us have the economic influence to change policies and investment decisions. How can ecologists influence that economically more powerful audience? One way may be through sharing the monetary value of ecosystem services, and showing how their loss may have serious economic consequences. The World Wide Fund for Nature estimates the contribution of nature to the world's economies to be US$125 trillion. Calculating the monetary value of individual resources, such as Australia's Great Barrier Reef (valued at US$5.7 billion per annum, and directly supporting 69,000 jobs), allows politicians and investors to see what may be lost if short-term commercial gain is the only driver of economic decision-making. However, there are risks associated with placing a price on nature. For example, it was suggested that fortunes will be made from discovering new drugs in rainforest plants. But modern pharmacology relies less on bioprospecting and more on computer modelling and laboratory synthesis to develop new drugs. Does this make rainforests now worth less in the eyes of investors? Assessing the monetary value alone is problematic.

RELATED TOPICS
See also
ECOSYSTEM SERVICES
page 118

POLLINATION & SOCIETY
page 122

OVERHARVESTING
page 128

3-SECOND SURVEY
Not everyone wants to save biodiversity for its own sake; placing a cash value on the benefits gained from nature may help persuade people.

3-MINUTE STUDY
Blue whales breed slowly and are rare, so any sustainable harvest would be small. Mathematician Colin Clark observed that rather than sustainably harvesting blue whales, from a purely financial perspective it would be better to kill them all and invest the money in industries with a better rate of return. Clark argued that this showed that decision making should not just be about profit; species' worth should be measured by more than their monetary benefits.

3-SECOND BIOGRAPHIES
COLIN W. CLARK
1931–
Canadian mathematician who laid the foundations of ecological economics.

ROBERT CONSTANZA
1950–
American academic who helped found the field of ecological economics.

30-SECOND TEXT
Mark Fellowes

Will putting a price tag on nature's contributions to global societies and economies help us value it more?

ECOSYSTEM SERVICES

the 30-second treatment

Ecosystem services are the benefits provided to humanity by ecosystems. We rely upon plants for breathable air, pollinators and productive soils for our food and drinking water, and forests for materials. While this was a given for ecologists, it was only in 2005 that global policymakers started to explicitly consider the economic consequences of ecosystem services. The United Nations' Millennium Ecosystem Assessment report introduced the term 'ecosystem services' to the wider public. The report noted that the degradation of ecosystems was occurring at a faster pace than at any other time in human history, causing huge and irreversible biodiversity losses. As the world has been changed to produce our food and materials, huge damage has been caused to four key areas: in supporting services, including nutrient cycling and pollination; provisioning services, which are the products of the natural world, from hydroelectric power to food, fibre and genetic resources for crop improvement; regulation services, such as natural pest control, water purification and carbon sequestration; and cultural services, including for ecotourism, cultural and spiritual benefits. To maintain ecosystem services, there needs to be a closer link between ecologists, economists and policymakers, people whose perspectives have traditionally not been closely aligned.

RELATED TOPICS
See also
VALUING BIODIVERSITY
page 116

POLLINATION & SOCIETY
page 122

THE SIXTH MASS EXTINCTION
page 138

3-SECOND BIOGRAPHY
GRETCHEN DAILY
1964–
American ecologist and proponent of the concept of ecosystem services; co-founded the Natural Capital Project, which aims to incorporate environmental issues into business practice and public policy.

30-SECOND TEXT
Mark Fellowes

3-SECOND SURVEY
Ecosystem services keep our species alive; at the current rate of damage their benefits and our quality of life will be significantly eroded.

3-MINUTE STUDY
Nature-based solutions is an approach that works with the grain of nature, to help build the benefits of ecosystem services into planning. In many cities, projects to develop green infrastructure, such as vegetated roofs and tree planting, provide ecosystem services through reducing air pollution, flood risk and urban heat stress, while boosting biodiversity and providing improved recreational and aesthetic benefits. By developing a 'green economy', nature-based solutions help create employment and economic growth.

Exposure to trees and parks improves people's mental and physical health; green infrastructure could transform cities and city life.

SOIL ECOLOGY

the 30-second treatment

3-SECOND SURVEY
While hidden, ecological interactions in the soil determine global patterns of plant diversity, agricultural productivity and even how the Earth will respond to climate change.

3-MINUTE STUDY
Ectomycorrhizal fungi (EMF) act in a similar manner to AMF, but are more commonly found in symbiotic relationships with trees. Recent work has shown that trees also use EMF networks to share nutrients between each other and even with other species. Even more extraordinary is the discovery that trees use EMF networks to share information in the form of hormones and defence chemicals. Trees are talking to each other, silently helping and warning their neighbours.

Soil is a product of its bedrock, with material eroded and changed through physical and chemical processes over time, and fertilized and strengthened by soil flora and fauna. There is more life in the soil than in any other environment. While insects, earthworms and moles may spring to mind as typical soil dwellers, life in the soil is predominantly microscopic. A gram of soil may hold up to 3 million bacteria. Soil biodiversity helps regulate the global carbon cycle. Soils contain more carbon than plants and the atmosphere combined, and the amount stored is partly determined by changes in the abundance and diversity of soil micro-organisms. Arbuscular mycorrhizal fungi (AMF) are found in the roots of 80 per cent of vascular plants, where they live in symbiosis. The AMF and plant exchange nutrients, improving plant health and growth through the provision of phosphorus and other scarce micronutrients. AMF also exude glomalin, a glycoprotein that may help sequester considerable volumes of carbon. Unfortunately, most modern crop production practices damage this relationship. The use of fertilizers and fungicides, combined with mechanical tilling of the soil, reduce the abundance of AMF. Agricultural approaches need to be adopted that minimize erosion, and improve soil fertility and carbon storage. While more costly, the long-term benefits are clear.

RELATED TOPIC
See also
ECOSYSTEM SERVICES
page 118

3-SECOND BIOGRAPHIES
ARISTOTLE
384–322 BCE
Greek philosopher who understood the role of earthworms, calling them the 'intestines of the Earth'.

SARA F. WRIGHT
fl. 1996–
American scientist who discovered glomalin, a product of AMF that may store one-third of the world's soil carbon.

SUZANNE SIMARD
fl. 1997–
Canadian ecologist who showed that trees can communicate with other trees through shared mycorrhizal networks.

30-SECOND TEXT
Mark Fellowes

A healthy soil ecosystem maintains productivity.

POLLINATION & SOCIETY

the 30-second treatment

3-SECOND SURVEY

We rely on pollinators for many of humankind's most important crops, but modern agricultural techniques are a leading cause of their decline.

3-MINUTE STUDY

Not all pollinators are equal. While domesticated honey bees and a variety of wild bees deposit the same amount of pollen on strawberry flowers, strawberries resulting from visits by wild bees were on average over two-thirds larger than those pollinated by domesticated honey bees. Indeed, for many crops, a species-rich community of pollinators is associated with increased productivity. Simply replacing wild pollinators with domesticated ones will not be the solution to pollinator declines.

Declines in pollinator numbers have been recorded in every continent, apart from Antarctica. Without pollinators, agricultural productivity plummets; one-third of global food production comes from insect-pollinated crops. Most horticultural, orchard and many forage crops rely upon pollination, mainly from insects, to produce the volume of outputs society demands. In Europe, 84 per cent of crop species directly depend on insect pollinators, and across the world, some 70 per cent of our 124 primary crop species require pollinators. Populations of wild pollinators have been devastated through careless pesticide use, habitat destruction, climate change and the introduction of parasites and diseases. Moreover, many crops are grown in monocultures, so pollinators cannot survive outside of the crop's flowering season due to a lack of alternative resources. As a result, agriculture increasingly relies on domesticated pollinators. In North America, honey bees are transported across the continent in huge trucks to ensure commercially viable crop production. For example, the survival of California's almond industry is predicated on farmers importing almost half of the honey bees in the United States, in around a million hives, for successful pollination. Changing agricultural practices to avoid monoculture and extensive pesticide use could help restore pollinator numbers.

RELATED TOPIC
See also
ECOSYSTEM SERVICES
page 118

3-SECOND BIOGRAPHIES
DAVE GOULSON
1965–
British entomologist and author who was one of the first to show that neonicotinoid insecticides were associated with bee declines.

MARYAM HENEIN
fl. 1998–
Canadian investigative journalist whose documentary 'Vanishing of the Bees' (2009) helped bring pollinator declines to public attention.

30-SECOND TEXT
Mark Fellowes

Pollinator decline is a massive threat to agriculture; stemming the loss requires increased habitat suitability and reduced pesticide use.

BIOLOGICAL CONTROL

the 30-second treatment

Pders are as old as agriculture.

Pests are as old as agriculture.
Humankind's approach to controlling pests typically relies on chemical pesticides, but if unwisely used these can have dreadful ecological consequences, such as the huge population declines seen in birds of prey caused by the overuse of the insecticide DDT. But there are alternatives. Predators, parasitoids and pathogens can be used to control pests. Predators such as cats were domesticated to control rodents, but as predators tend to be generalist in their prey choice, they can cause more problems than they cure by killing non-target species. Parasitoids and pathogens can be much more host-specific. Parasitoids are typically small wasps or flies that lay their eggs in or on other insects, where the eggs hatch and the larvae consume their host, killing it. The adults emerge and hunt down their next host. Bacterial, fungal and viral pathogens can also be host-specific, and some can be applied like a chemical insecticide by spray when needed. Biological control can be exceptionally effective. The Californian citrus industry was almost destroyed by the invasive Australian cottony cushion scale in the late nineteenth century. A parasitoid fly and a predatory beetle were introduced from the pest's original Australian range. Together they controlled the scale insect, saving farmers' livelihoods.

3-SECOND SURVEY
Agriculture relies on our ability to control pests; natural enemies provide the best hope of doing this without harming non-target species.

3-MINUTE STUDY
A small moth called *Cactoblastis cactorum* illustrates the benefits and risks of using biological control. A species of prickly pear, a South American cactus, was introduced to Australia for fencing purposes. Lacking enemies, it rapidly spread, overwhelming over 40,000 km² (15,400 square miles) of productive farmland. *Cactoblastis*, a prickly pear herbivore from Argentina, brought the cactus under control. But now *Cactoblastis* is spreading around the world, and today threatens native prickly pear species in North America.

RELATED TOPIC
See also
. . . AND BEING EATEN
page 62

3-SECOND BIOGRAPHIES
JI HAN
263–307
Botanist who wrote the first record of biological control, involving weaver ants placed in citrus plantations to control leaf-eating insects.

CHARLES VALENTINE RILEY
1843–95
The 'father of biological control' who worked on cottony cushion scale and saved the French wine industry from *Phylloxera*.

HANS RUDOLF HERREN
1947–
Led the fight against the cassava mealybug using parasitoids, preventing the loss of one of Africa's most important crops.

30-SECOND TEXT
Mark Fellowes

Millions of farmers across the world rely on the natural enemies of pests to protect crops and livestock.

VECTORS OF DISEASE

the 30-second treatment

Vectors are organisms that can transmit diseases between species. For humans, the most prevalent vectors are blood-feeding insects and ticks, which take in disease-causing parasites, bacteria and viruses as they feed on the blood of an infected person. They are then transmitted to a new host when the vector takes its next blood meal. It is estimated that 17 per cent of all infectious diseases are transmitted by vectors. Over 400,000 people die of mosquito-borne malaria each year, with young children most at risk. This is increasingly a problem in towns and cities, which combine plentiful breeding grounds with high densities of people to feed from. The urban heat-island effect means that these areas are warmer than their surroundings. Insect development is affected by ambient temperature, and a 0.50°C (1°F) increase in temperature in sub-tropical areas can increase mosquito numbers by 30 to 100 per cent. While insecticides provide the first line of defence, the rapid spread of resistance is limiting their efficacy. Biological control provides an alternative. Fish and predatory insects can reduce numbers of mosquito larvae if added to water butts, and highly specific fungal and bacterial pathogens can also be used. The burden of vector-borne diseases largely falls on tropical countries, where methods of sanitation and vector control are limited.

3-SECOND SURVEY
Vectors do not cause disease themselves, but because of their behaviour they help diseases spread from one host to another in a highly efficient way.

3-MINUTE STUDY
While people living in tropical and sub-tropical regions are most at risk of vector-borne diseases, those living in temperate regions do not escape. Tick-borne Lyme disease is the most common vector-borne disease in North America and Europe, infecting tens of thousands of people each year. Humans are an accidental host, as the ticks typically feed on deer and rodents. Healthy ecosystems, with plenty of natural rodent predators, may be the best defence against Lyme disease.

RELATED TOPICS
See also
PARASITES
page 64

TROPICAL FORESTS & SAVANNA
page 106

3-SECOND BIOGRAPHIES
SIR RONALD ROSS
1857–1932
British doctor awarded the Nobel Prize for showing that mosquitoes transmitted malaria.

WILHELM BURGDORFER
1932–2014
Swiss-born American parasitologist who discovered the bacteria that causes Lyme disease.

30-SECOND TEXT
Mark Fellowes

Vectors are the keystone in the chain of infection of many important diseases; ecological change is increasing the risk of transmission.

OVERHARVESTING

the 30-second treatment

Harvesting is the collection

and use of living resources, whether wild or cultivated. Unfortunately, history shows that overharvesting, the unsustainable use of wild animals and plants, has had irreversible consequences for biodiversity. The concept of sustainability relies upon the idea of not taking more than can be replaced by reproduction. A species' population size is limited by its habitat's carrying capacity, determined by the availability of resources such as food and water. At carrying capacity, removing individuals has little effect on a species' abundance, as individuals are replaced by others that would otherwise not have survived. If too many are removed too quickly, however, then populations will decline. Understanding the rate and timing of replacement allows ecologists to estimate the point at which overharvesting occurs. Unfortunately, such ecological insights are rarely used. For example, early whalers drove North Atlantic right whales and bowhead whales to near extinction by the nineteenth century. Over time, technology developed, each innovation allowing new species to be exploited, and whalers moved from grey whales to humpbacks, sperm whales to blue whales, driving each to near extinction. Despite the overwhelming evidence for losses, it took decades of debate to halt the hunting. This illustrates the challenges faced by ecologists.

RELATED TOPIC
See also
THE SIXTH MASS EXTINCTION
page 138

3-SECOND SURVEY
Humans must harvest natural resources; while ecology shows how this can be done sustainably, human society seems to act against this.

3-MINUTE STUDY
Sea otters were almost exterminated from North America due to hunting. Sea otters eat sea urchins, and in turn sea urchins eat kelp. The loss of otters allowed sea urchin populations to rise, which then removed kelp from huge areas. Kelp is key to the maintenance of healthy fish stocks, so removing sea otters caused reductions in the populations of many other species. Species exist in ecological networks; overharvesting one species can have enormous ecological consequences.

3-SECOND BIOGRAPHIES
ROGER PAYNE
1935–
American environmentalist who discovered whale song and became an advocate for whale conservation.

SIR JOHN BEDDINGTON
1945–
British ecologist whose work on sustainable harvesting influenced policymakers across the world.

30-SECOND TEXT
Mark Fellowes

Whaling illustrates how science can provide data, but the prevention of irreversible loss necessitates political will, responding to public demands.

CONSERVATION BIOLOGY

the 30-second treatment

3-SECOND SURVEY
While individual species receive the most attention, conservation biology aims to protect biodiversity at all levels, from genes to ecosystems.

3-MINUTE STUDY
The International Union for the Conservation of Nature's Red List is a barometer of threats to species. Of over 105,000 species assessed (2019), 27 per cent are threatened with extinction, including 25 per cent of mammal, 40 per cent of amphibian and 33 per cent of coral species. While we understand the threats to well-known groups, there are major knowledge gaps for taxa as biodiverse as fungi and invertebrates. We can only protect what we know; the inventory of life on Earth needs completing.

Conservation biology is the science of protecting and enhancing populations of rare and declining species. Declining numbers are commonly caused by habitat loss and overexploitation, leading to a spiral of decline as populations become smaller and fragmented, exposing the species to disease, inbreeding, genetic drift and chance demographic effects. This is the extinction vortex, where one cause of decline leads to another in a spiral effect. Conservation biology helps species escape the vortex in two ways. *In situ* conservation relies upon protecting habitats and reducing threats. *Ex situ* conservation involves captive breeding, with the aim of later releasing individuals into suitable habitats. The California condor illustrates how these approaches work together. By 1987, the condor was extinct in the wild, with the last 22 individuals held in captivity. In the 1990s, breeding in zoos allowed their reintroduction into protected areas of California, Arizona and Mexico. In 2015, more condors were born in the wild than died that year, and the wild population has now passed 300. Captive breeding continues, augmenting wild populations, and condors are protected in the wild from major threats. The California condor was once seen as a lost cause, but this magnificent species' continued existence in the wild shows that conservation works.

RELATED TOPIC
See also
THE SIXTH MASS EXTINCTION
page 138

3-SECOND BIOGRAPHIES
SIR DAVID ATTENBOROUGH
1926–
Influential British broadcaster who has helped bring conservation issues to the world's attention.

MICHAEL E. SOULÉ
1936–
American ecologist who helped coin the term 'conservation biology' in 1978.

30-SECOND TEXT
Mark Fellowes

The California condor is an example of how conservation can prevent extinction, but also how prevention is better than cure.

27 May 1907
Born near Springdale,
Pennsylvania, USA

1929
Graduates with honours
in Biology, Pennsylvania
College for Women

1932
Graduates with master's
degree in Zoology from
Johns Hopkins University

1935
Takes a temporary role
as copywriter for a US
Bureau of Fisheries radio
programme

1936
Employed full-time by the
US Bureau of Fisheries as
a junior aquatic biologist

1949
Appointed Chief Editor
of Publications for the US
Fish and Wildlife Service

1951
Publishes *The Sea Around
Us*, which wins the
National Book Award
for Non-fiction

1955
Publishes *The Edge of
the Sea*

1960
Diagnosed with breast
cancer

1962
Publishes *Silent Spring*

14 April 1964
Dies of cancer
complications at home in
Silver Spring, Maryland

RACHEL CARSON

Rachel Carson was born in 1907, and grew up on a farm in rural Pennsylvania. She developed a love of nature at a young age, encouraged by her mother. She studied Biology at college, continuing to a master's degree in Zoology at Johns Hopkins University in Maryland. She was the first woman to take and pass the US Federal civil service test and on her full-time appointment as junior biologist in 1936, she was only the second woman permanently employed by the US Bureau of Fisheries, one of the precursors of the US Fish and Wildlife Service. She later rose to become chief editor of all its publications in 1949.

Carson was an expert in marine biology, and published many books on sea life, most notably *The Sea Around Us* (1951), which sold more than 250,000 copies. But it was the ground-breaking *Silent Spring* (1962) that became one of the foundations of the modern environmental movement. The mass use of synthetic insecticides such as DDT started in the 1950s. Control over their application was lax, and it was not long before it became apparent that these chemicals also killed many other species. At the request of the National Audubon Society, Carson

interviewed farmers, doctors, scientists and wildlife experts, which resulted in *Silent Spring*.

This was a call to arms. Rather than being termed pesticides, Carson argued that the term biocide was better, because of the non-target organisms they killed. The chemical industry did not stand by. She was accused of gross exaggeration, threatened with lawsuits and accused of being a communist. President John F. Kennedy ordered an enquiry into her claims. This vindicated her, and led to increased control over pesticide use and ultimately the banning of DDT in the US and later around the world for most uses.

Carson died two years after the publication of *Silent Spring*. She had been diagnosed with breast cancer while writing the book. Selling more than 50 million copies in the US alone, and translated into over 30 languages, *Silent Spring* started a revolution in environmental awareness. It still resonates today. In 2012 it was designated a National Historic Chemical Landmark by the American Chemical Society due to its influence on environmentalism. Carson herself is memorialized in the name of research vessels, buildings, schools and prizes.

Mark Fellowes

ECOLOGY IN A CHANGING WORLD

ECOLOGY IN A CHANGING WORLD
GLOSSARY

anthropogenic A human activity that causes damage to the environment.

biodiversity All of the microorganisms, plants, animals, and the genetic diversity within them, as well as the diversity of habitats and ecosystems.

biodiversity crisis Species extinction is a natural process, but human action is driving the extinction of many more species than would be found naturally.

biosphere The biosphere contains all of planet Earth's ecosystems.

biotic homogenization Where ecological communities become increasingly similar over time due to species invasions.

bovine tuberculosis Caused by the bacterium *Mycobacterium bovis*, this infectious disease infects cattle, but also many other mammals including dogs, cats, humans and badgers.

captive breeding Taking animals into captivity and breeding them, in the hope that they can later be released back into the wild.

ecosystem A community of organisms interacting with each other and their physical environment.

ecotourism A type of tourism in which people visit natural environments, and the income generated contributes towards the local economy and conservation efforts.

feedback mechanism (climate change) Complex, and often unpredictable loop, where one factor affects the next in a vicious circle that can accelerate climate change.

fossil fuel The remains of long-dead plants and animals, including coal, natural gas and petroleum, which are used for fuel and the production of goods.

keystone species A species that plays a vital role within the ecosystem; its removal would cause dramatic differences in how that ecosystem functions.

monoculture A cultivated area with a single crop species grown.

pollination service Many plants rely on animals, such as bees, to transfer their pollen from the male anther to a female stigma of the same or another flower of the same plant or species, enabling fertilization and seed production.

pollinator An animal that moves pollen from one plant to another. Many pollinators are insects (such as bees, moths and beetles), but many birds and mammals also act as pollinators.

rabies An infectious virus that is deadly to humans and other animals. It is normally transmitted via saliva through a bite.

radio telemetry Technology used to track an animal's location using radio signals. The radio transmitter is attached to an animal and a scientist uses an antenna to pick up the radio signal. Scientists locate the animal at regular intervals and can then build a picture of where that animal spends its time.

renewable energy Wind, waves, tides, sunlight and geothermal heat all produce energy that do not become depleted over time.

tipping point (climate change) As global temperature rises, scientists believe that a threshold will be reached from which it will be impossible to recover. As an example, if carbon reaches over 1,200 parts per million, scientists believe that there would be reduced cloud cover over the oceans, triggering further rises in global temperature.

THE SIXTH MASS EXTINCTION

the 30-second treatment

3-SECOND SURVEY
Humans are causing the loss of biodiversity at an unprecedented rate, and if this decline is not reversed, it will threaten our own survival.

3-MINUTE STUDY
Many species are being lost, even from protected areas. Reducing or even reversing this decline is some way off, but action is being taken. Captive breeding and reintroduction have seen some successes. The Mauritius kestrel, for example, was once the rarest bird in the world, reduced to just four birds, yet conservation efforts using captive breeding have increased the population to around 350.

No species lasts forever. Our planet has undergone five mass extinction events in its geological past, where up to 90 per cent of species were lost each time. Today we are facing the sixth such extinction, caused by human activity. Habitat destruction, persecution and the changing climate are some of the reasons that so many species are being lost. One lost species is the passenger pigeon: once endemic to the deciduous forests of North America, passenger pigeons may have numbered as many as 5 billion. Yet, after extensive hunting in the nineteenth century, the last captive pigeon, Martha, died in Cincinnati Zoo in 1914. While the loss of charismatic species such as rhinos is notable, many species are lost before scientists have a chance to identify and classify them. Each species has its place in an ecosystem. Failing to reverse declines threatens the survival of humankind, with ecosystem function jeopardized. For example, pollinating insects are declining rapidly in number. This decline is mostly due to agricultural practices, which include spraying chemicals and creating monocultures in which pollinators cannot survive. We rely on these insects to pollinate 75 per cent of food crops, and a lack of pollination services would be a disaster for food production, which would affect developing countries the most.

RELATED TOPICS
See also
COUNTING SPECIES
page 18

ECOSYSTEM SERVICES
page 118

CONSERVATION BIOLOGY
page 130

3-SECOND BIOGRAPHY
CARL GWYNFE JONES
1954–
Conservation biologist at the Durrell Wildlife Conservation Trust who saved the Mauritius kestrel from extinction.

30-SECOND TEXT
Becky Thomas

Human actions, such as hunting and using insecticides on crops are causing declines in many species.

CLIMATE CHANGE

the 30-second treatment

RELATED TOPIC
See also
HUMAN–WILDLIFE CONFLICT
page 144

3-SECOND BIOGRAPHIES
DR HOESUNG LEE
1945–
Chair of the Intergovernmental
Panel on Climate Change
since 2015, which supplies
governments and the
public with the scientific
understanding behind
climate change.

ALBERT GORE JR
1948–
US Vice President 1993–2001,
who won the Nobel Peace
Prize (joint with the IPCC) in
2007 for his work in climate
change activism.

30-SECOND TEXT
Becky Thomas

The biggest threat to biodiversity and human well-being is anthropogenic climate change. Human activity is causing the planet to warm up, creating many knock-on effects, such as extreme weather events. Increases in carbon dioxide levels in the atmosphere since the Industrial Revolution are caused by activities such as the burning of fossil fuels, deforestation and agriculture. When fossil fuels are used, the carbon stored within them is released into the atmosphere as carbon dioxide. Carbon dioxide and other gases, such as methane, trap heat within the atmosphere, causing global warming. As a result, the oceans are heating up, causing coral bleaching across the world, and the Greenland and Antarctic ice sheets are decreasing in size at an alarming rate. Species need to adapt to these rapid changes, but many are already feeling the heat. Polar bears depend upon sea ice to live and are vulnerable to the changing climate. With their slow growth and reproductive rates, they are less able to adapt quickly to changes in their environment. Governments and people globally need to work together to reduce the threat faced by climate change before it is too late. People can take action, for example, in cutting back consumption of red meat and reducing numbers of flights, to using energy suppliers that supply energy from renewable sources.

3-SECOND SURVEY
The Earth's climate is changing due to human action, devastating habitats and the species within them. Unless we make drastic changes, species will have to adapt or they will perish.

3-MINUTE STUDY
Scientists talk about the dangers of the world's temperature increasing 1.5°C (2.7°F) above pre-industrial levels, but it is difficult to predict the effects of feedback mechanisms or tipping points – chains of events that once started are difficult to stop. The ice caps at each pole help to cool the planet, reflecting sunlight back, but as our climate warms, the ice melts at an increasing rate. Reduced ice decreases this cooling mechanism, triggering further warming in a feedback loop that would be hard to control.

Our warming climate is already negatively affecting many species, especially those living in the polar regions.

URBAN ECOSYSTEMS

the 30-second treatment

Cities and towns are our newest habitat type and are sculpted and created to suit our needs. With over half the world's population now living in urbanized areas, some species are pushed out, while many others adapt and thrive in this artificial world. In a process known as biotic homogenization, the species found city-to-city look surprisingly similar, even if those cities are on different continents. Urban dwellers make choices about what to plant, and often choose the same types of trees and plants. People have also accidentally and deliberately taken all sorts of species with them as they travel across the globe, such as the European starling and house sparrow, which can now be found in cities across the world. Those that thrive in urban ecosystems are generalist species, meaning that they eat a wide range of foods and benefit from the structures that people create. One surprising group benefiting from urban habitats is birds of prey, such as falcons, which enjoy feasting on urban pigeons and nesting on tall buildings. Cities present opportunities for some species, but as urban habitats expand, they are creating barriers, and people need to work harder to make these manufactured habitats more accessible to other species.

3-SECOND SURVEY
Urban ecosystems create barriers for many species, but opportunities for those able to exploit the novel habitats that they provide.

3-MINUTE STUDY
Within towns and cities, there are opportunities to create spaces that can be shared with wild species. Gardens and spaces around people's homes are places that they control and manage, and creating a space for wildlife can be surprisingly easy. In some countries, up to 50 per cent of homeowners feed wild birds and many also choose to put up nesting boxes for birds and bats. Making changes to the nature of the spaces around us not only benefits wildlife, but also our own health and well-being.

RELATED TOPICS
See also
HABITATS & NICHES
page 22

NATURE-DEFICIT DISORDER
page 148

3-SECOND BIOGRAPHY
LUKE HOWARD
1772–1864
Chemist and meteorologist who first discovered that urban areas are warmer than the surrounding countryside. We now call this the 'urban heat island effect'.

30-SECOND TEXT
Becky Thomas

Cities can be hostile places for wild animals, but some species have learnt to exploit the resources available.

HUMAN–WILDLIFE CONFLICT

the 30-second treatment

Many species are in population decline because of negative interactions with humans. Whether through direct interactions such as hunting or persecution, or indirectly, via the destruction of habitats, the growing human population is putting pressure on nature. Humans change the landscape like no other species, and conflicts can and do occur over space and food. Large numbers of humans live in close proximity to predators, such as lions, or to species with whom they may compete for food, such as elephants. Imagine living in a city where you know that a leopard could be lurking nearby. This is the reality that many people face, especially in countries such as India. Leopards are losing natural habitat through urbanization, and their prey are being reduced in abundance. They can be drawn to cities because of the lure of an easy meal, as there are high numbers of stray dogs. There are, however, benefits to humans in living alongside leopards. In India, dog bites are common, and transmission of rabies to people can occur. Those living in areas with high leopard numbers are less likely to suffer dog bites, yet the risks that leopards pose to people are still high. With human populations set to continue to rise, conflicts with wildlife are expected to increase, and they will need to be managed carefully.

RELATED TOPIC
See also
URBAN ECOSYSTEMS
page 142

3-SECOND BIOGRAPHY
ROSIE WOODROFFE
fl. 1993–
Scientist working on reducing conflicts between farmers and badgers, since badgers have been implicated in the rise of TB in cattle.

30-SECOND TEXT
Becky Thomas

3-SECOND SURVEY
Human populations are increasingly coming into conflict with wildlife, and these conflicts can be complex and difficult to manage, leading to the persecution of many species.

3-MINUTE STUDY
Human–wildlife conflicts are common and can range from property damage to livestock predation, but there are many novel ways to reduce this conflict. Electric fences have been used in India to prevent crop raiding by elephants, while ecotourism is being adopted in many countries to boost local economies and to improve the value that people attribute to those species that cause conflict.

With human populations booming, people are encroaching on natural habitats and creating rivalry with species such as leopards.

3 April 1955
Born in Chennai, India

1977
Receives Bachelor of
Science degree from the
University of Madras

1985
Completes PhD at
the Indian Institute
of Science

1986–
Becomes Professor at
the Centre for Ecological
Sciences, Indian Institute
of Science and
establishes Nilgiri
Biosphere Reserve

1991
Becomes a Fulbright
Fellow at Princeton
University

1993–2004
Joins the Project
Elephant Steering
Committee (and chairs
it from 1997)

1994
Publishes the book
*Elephant Days and
Nights: Ten Years with
the Indian Elephant*

1997
Helps to establish
the Asian Nature
Conservation Foundation

2000
Becomes Fellow of
the Indian Academy
of Sciences

2003
Awarded the Whitley
Gold Award for
International Nature
Conservation

2006
Awarded the
International Cosmos
Prize, recognizing
research that promotes
the coexistence between
nature and mankind

RAMAN SUKUMAR

Being raised in the sprawling city of Chennai, India, in no way hindered Dr Sukumar's fascination with the natural world; in fact, it served to steer his life's passions and work. Called 'vanavasi' (Tamil for 'forest dweller') by his grandmother, at around the age of 15, Sukumar became interested in conflicts between humans and wildlife, as well as developing a concern for the loss and conservation of the wild species around him. His first degrees from the University of Madras, in 1977 and 1979, were in Botany, but he began his doctoral thesis at the Indian Institute of Science looking at conflicts between humans and Asian elephants.

Sukumar set about to understand these interactions and how they can be managed. Technology has been an important part of his work, using radio telemetry to identify the habitats of the elephants as well as understanding more about the corridors that elephants use when they travel between nature reserves. In 1986, Sukumar became a professor at the Centre for Ecological Sciences, attached to the Indian Institute of Science.

His passion for conservation and his local area led to Sukumar's involvement in the development of the Nilgiri Biosphere Reserve in 1986, an international biosphere reserve in the Western Ghats. Within the reserve he expanded his research by looking at the way that the past climate has shaped the vegetation present today. Using this information has allowed him to make predictions about how future climate change could affect the ecology of the region.

Sukumar's conservation and work with elephants continued, he advised the Indian government through the Project Elephant Steering Committee from 1993 to 2004, publishing a number of notable texts and he set up the Asian Nature Conservation Foundation in 1997. The foundation aims to reduce conflicts between people and elephants, and, in particular, to deter crop-raiding elephants.

The extensive international accolades awarded to Sukumar demonstrate his contribution to wildlife conservation and in reducing conflicts between people and nature. He was the first Indian scientist to win the International Cosmos Award, a prestigious prize in Ecology, and was commended by the Indian Prime Minister for his contributions to the 2007 Nobel Prize-winning Intergovernmental Panel on Climate Change (IPCC). Sukumar continues his work as a Professor at the Centre for Ecological Sciences, at the Indian Institute of Science in Bangalore.

Becky Thomas

NATURE-DEFICIT DISORDER

the 30-second treatment

The lifestyles of people have changed dramatically within the last century. In the past, most people lived on or near the land as farmers or hunters. Their lives were entwined with nature. Today, over 50 per cent of people worldwide live in urban areas, and humans are increasingly disconnected from the natural world. People of all ages, especially children, are spending less time outside and more time in front of screens, with implications for their health and well-being. This has been linked to attention disorders and depression in many developed nations, including the US, UK, Canada and Australia, and has been termed 'nature-deficit disorder'. Research has shown that the health benefits of being outside, or even being able to see nature through a window while recovering in a hospital, are vitally important. This disassociation from nature is also a big problem for nature conservation. As people become more detached from nature, their desire to protect the natural world around them diminishes. Humans depend on nature to provide food and other material needs, but there is also another need, linked to our own health and well-being, that is less well known and understood.

Young children and adults in developed countries are spending less time outdoors, and have little connection with nature.

REWILDING

the 30-second treatment

We have reached a biodiversity crisis, in which human activity is leading to the decline – and even extinction – of many species around the world. Ecologists use lots of different techniques to try to reduce this decline, which often involves actively managing a landscape, such as creating a particular type of habitat, to promote the conservation of one or more species. Rewilding takes a different approach: it is large in scale and has limited human intervention, letting nature take back control. Keystone species, which are species that have a disproportionately large and positive effect on their environment relative to their abundance, may be reintroduced. The aim is to create a system that needs little direct management and which benefits the biodiversity of the area. Rewilding projects are taking place all over the world, with one of the largest spanning from Yellowstone in the US to Yukon in Canada, stretching over 3,200 kilometres (2,000 miles). For many species, conservation needs to occur on this scale as many national parks are not big enough to protect species with large ranges, such as wolves and bears. These rewilding projects create habitat bridges between different protected areas, enabling self-sustaining populations to thrive.

RELATED TOPIC
See also
ECOSYSTEM SERVICES
page 118

3-SECOND SURVEY
Rewilding is a form of habitat restoration with reduced human intervention that promotes species conservation, often on a large scale.

3-MINUTE STUDY
When rewilding projects reintroduce lost species, positive and sometimes unexpected benefits can occur. Beavers have been reintroduced in many rewilding projects, substantially changing a habitat as they construct dams, creating wetlands vital for many other species. Many rewilding projects have reintroduced animals that feed in different ways: pigs rootle the ground, making space for wildflowers to grow; deer and wild horses browse, preventing trees from taking over and dominating.

3-SECOND BIOGRAPHY
FRANS VERA
1949–
Dutch ecologist involved in a large-scale rewilding project in the Oostvaardersplassen nature reserve in the Netherlands. His book *Grazing Ecology and Forest History* identifies the importance of grazing animals in the ecology of an area.

30-SECOND TEXT
Becky Thomas

Rewilding projects change a landscape, often through the reintroduction of species such as wild boar, which encourage other species to flourish.

APPENDICES

RESOURCES

BOOKS

Applied Ecology: Monitoring, managing, and conserving
Anne Goodenough and Adam Hart
Oxford University Press (2017)

The Beak of the Finch: A Story of Evolution in Our Time
Jonathan Weiner
Vintage (1995)

Demons in Eden: The Paradox of Plant Diversity
Jonathan Silvertown
University of Chicago Press (2008)

Ecology
William D. Bowman, Sally D. Hacker & Michael L. Cain
Oxford University Press (2018)

Ecology: The Economy of Nature
Robert Ricklefs & Rick Relyea
W.H. Freeman & Co (2014)

Feral: Rewilding the Land, Sea and Human Life
George Monbiot
Penguin (2014)

In the Shadow of Man
Jane Goodall
Weidenfeld & Nicolson (1999)

An Introduction to Behavioural Ecology
Nicholas B. Davies, John R. Krebs, Stuart A. West
Wiley-Blackwell (2012)

Last Child in the Woods: Saving Our Children from Nature-deficit Disorder
Richard Louv
Atlantic Books (2010)

Serendipity: An Ecologist's Quest to Understand Nature
James A. Estes
University of California Press (2016)

Silent Spring
Rachel Carson
Penguin Modern Classics (2000)

The Sixth Extinction
Elizabeth Kolbert
Bloomsbury Paperbacks (2015)

Trophic Cascades: Predators, Prey, and the Changing Dynamics of Nature
John Terborgh
Island Press (2010)

Why Big Fierce Animals Are Rare – An Ecologist's Perspective
Paul A. Colinvaux
Princeton University Press (1979)

Why Evolution is True
Jerry A. Coyne
Oxford University Press (2010)

Wilding: The Return of Nature to a British Farm
Isabella Tree
Picador (2019)

The Wolf's Tooth: Keystone Predators, Trophic Cascades, and Biodiversity
Cristina Eisenberg
Island Press (2011)

WEBSITES

International Society for Behavioural Ecology
www.behavecol.com

British Ecological Society
www.britishecologicalsociety.org

Ecological Society of America
www.esa.org

Cary Institute of Ecosystem Studies
www.caryinstitute.org

NOTES ON CONTRIBUTORS

CONSULTANTS

Mark Fellowes has been passionate about wildlife since an early age, spending his formative years in the wilds of the west of Ireland surveying birds and bringing bits of the countryside back home. He completed his BSc in Zoology and a PhD in Evolutionary Biology at Imperial College London, UK. Following a brief post-doctoral stint at the NERC Centre for Population Biology at Imperial, he joined the University of Reading as a lecturer. He became Professor of Ecology, and is now Pro-Vice Chancellor at the university. Professor Fellowes has published several books and numerous papers in research journals. His current work focuses on interactions between people and wildlife, with an emphasis on urban ecosystems. He works on insects, birds and mammals, with ongoing projects in Ghana, Nigeria, India, the US and Brazil, as well as in the UK. Professor Fellowes undertakes a range of outreach work to engage the public and media in understanding the importance of biodiversity and was listed as one of the UK's 100 most influential men under the age of 40 by *Esquire* magazine in April 2004.

Becky Thomas is an urban ecologist and Senior Teaching Fellow at Royal Holloway, University of London, UK. She has a BSc in Zoology from Royal Holloway University of London, and an MSc in Wildlife Management and Conservation and PhD in Conservation Ecology from the University of Reading. Her research interests focus on the conservation biology and ecology of birds and mammals, specifically in how people's decisions affect the ecology of wild species. Dr Thomas is particularly interested in how human activity affects ecological interactions at a range of scales, and especially in trying to uncover some of the unexpected and unpredicted consequences of our behaviour.

CONTRIBUTORS

Heather Campbell is a Lecturer in Entomology at Harper Adams University, UK. She completed her PhD in Ecology at the University of Reading and has since worked at universities in Australia and South Africa as an expert in biodiversity and the conservation of insects. Her research focuses on community ecology and how insects interact and respond to natural and man-made changes in the environment.

James Cook studied Zoology at the University of Oxford and then completed a PhD in Evolutionary Biology at Imperial College London, UK. He is now Professor of Entomology at Western Sydney University, Australia. His research focuses on insect ecology and insect/plant interactions, and in particular the biology of insect pollinators and pollination.

Julia Cooke is an Australian plant ecologist who studies how plants acquire, allocate and use resources, with a focus on plant functional traits and how plants use silicon. She is a Senior Lecturer at The Open University, UK.

Stephen Murphy is a plant community ecologist working at the Missouri Botanical Garden in St Louis, Missouri, USA. He earned his PhD in Evolution, Ecology and Organismal Biology from The Ohio State University. His research focuses on understanding the processes generating and maintaining patterns of biodiversity.

Sarah Papworth is a Senior Lecturer at Royal Holloway University of London, UK. She has broad research interests in conservation and behaviour, focusing on primate responses to human hunters and tourists, and human perceptions of the natural environment. She has a BA in Anthropology from the University of Durham, an MSc in Ecology, Evolution and Conservation, and a PhD in Conservation Behaviour from Imperial College London.

Adam Smith is a global change scientist at the Missouri Botanical Garden who does research and outreach on the myriad ways humans are impacting the biosphere. He uses statistical models to understand the impacts of climate change, urbanization, agriculture and invasive species on ecosystems. He earned his PhD from the University of California, Berkeley, and prior to that volunteered full-time for a non-profit organization and taught English in Japan.

INDEX

ACKNOWLEDGEMENTS

The publisher would like to thank the following for permission to reproduce copyright material:

Alamy Stock Photo/ Historic Images: 24; SOTK2011: 86

Dreamstime/ Galinasavina: 101; Patricio Hidalgo: 89; Valentyna Chukhlyebova: 19

Flickr/ Biodiversity Heritage library: 15, 17, 19, 23, 35, 41, 49, 53, 62, 65, 75, 97, 99, 105, 111, 117, 129, 139, 143

Getty Images/ Vicki Jauron, Babylon and Beyond Photography: 23

The Graphics Fairy: 15, 39, 51, 151

Internet Archive/ University of Massachusetts Amherst Libraries: 61; Missouri Botanical Garden: 123

Jack Brett: 39, 107, 145

Library of Congress: 15, 39, 139

New York Public Library Digital Collections: 23, 49, 127, 129, 131

Sarah Blaffer Hrdy/Anthro-Photo: 42

Sarah Skeate: 90

Shutterstock/ acarapi: 107; akiyoko: 139; Alberto Zamorano: 119; Iditiya Rakasiwi: 86; aleks1949: 103; Alex Helin: 127; ALEXEY GRIGOREV: 97; Alexey Seafarer: 117, 141; Alin Brotea: 85; Alison Hancock: 119; Allahfoto: 127; Allween: 99; Amelia Martin: 125; Amit Kumar Photography: 145; Andrey Pavlov: 62; Andrzej Kubik: 107; Anna Kaminska: 139; annwizard: 37; Anton Balazh: 17; Arcady: 85; ArtHeart: 59; Artiste2d3d: 101; Ase: 99; BGSmith: 103; BigRoloImages: 141; bluehand: 69; Bodor Tivadar: 86; Brian A Wolf: 131; bwise: 119; Cat Downie: 111; Catherine Eckert: 125; Chichimaru: 77; Choksawatdikorn: 57, 99; chrisdorney: 149; clarst5; Claudia Schmidt: 65; Claudio Divizia: 145; COULANGES: 59; danm12: 97; David Martinez Perales: 141; David Osborn: 31; Delbars: 107; Deviney Designs: 82; dibrova: 77; Dirk M. de Boer: 33; dive-hive: 81; dkidpix: 53; Dreamframer: 123; Dubova: 89; dugdax: 51, 82; EA Given: 62; Elzbieta Sekowska: 65; Eric Isselee: 61, 107, 121, 151; Everett Collection: 65, 117, 149; Everett Historical: 139; Evlakhov Valeriy: 101; Exclusively: 86; eyeCatchLight Photography: 103; fotosutra: 81; frank60: 65; frank60: 143; FreeProd33: 123; Ga_Na: 149; gan chaonan: 86; ggw: 99; GoodStudio: 41; grafvision: 19; Gregory A.

Pozhvanov: 103; guraydere: 37; Hannah Prewitt: 69; Hans C. Schrodter: 77; Hayley Crews: 15; Hein Nouwens: 41, 75, 86, 89; Henrik Larsson: 85; Hesti Lestari: 97; Hyejin Kang: 49; I WALL: 141; Ihnatovich Maryia: 79; ImagesofIndia: 145; IrinaK: 99; J. Helgason: 121; jan j. photography: 62; jaroslava V: 103; javarman: 61; Jiri Foltyn: 119; johnfoto18: 99; johnfoto18: 99; Josh Anon: 33; justyean: 143; Kateryna Kon: 127; KellyNelson: 61; Ken StockPhoto: 119; kesipun: 123; Khadi Ganiev: 77; Kim Worrell: 131; Kletr: 81; kojihirano: 61; kovalto1: 105; KPixMining: 127; kwanchai.c: 57; Lane V. Erickson: 67; Leigh Richardson: 149; Lena_ Graphic: 86; Leonardo Gonzalez: 81; LeonP: 67; Lucas T. Jahn: 59; m. jrn: 123; Maksym Bondarchuk: 51; Manfred Ruckszio: 82; Manu M Nair: 111; Mariko Yuki: 31; Mario Pantelic: 61; MARIOS THEOLOGIS: 121; Marius Dobilas: 59; Mark Herreid: 62; mart: 86; Marzolino: 31, 49; MikeDotta: 79; MindStorm: 31; More Images: 41; Morphart Creation: 143; Mr Doomits: 89; MZPHOTO.CZ: 33; N1chEZ: 119; Nadiia_foto: 143; Nadine Pfeiffer: 59; narkorn: 123; NASI: 101; natacabo: 59; Nenov Brothers Images: 105; Nerthuz: 81; Netta Arobas: 141; New Africa: 119; nexusby: 77; Nor Gal: 67; Olga Vasilyeva: 145; Perfect Lazybones: 125; Peter Gudella: 103; Phichai: 127; Philip Schubert: 59; PhilipYb Studio: 33; photomaster: 31, 79; Photopunk: 86; Picture Partners: 62; PLotulitStocker: 51; Potapov Alexander: 85; Production Perig: 119; Protasov AN: 123, 139; Quick Shot: 107; Radiokafka: 145; Rafael Trafaniuc: 149; rangizzz: 117; Rastislav Sedlak SK: 119; Rattiya Thongdumhyu: 121; Rednex: 67; river34: 111; robert_s: 117; Roman King: 51; Rudmer Zwerver: 37; Rudra Narayan Mitra: 57; Runa0410: 79; Ryan M. Bolton: 67; Sam Dcruz: 59; SARANRUT PATCHARATANYANON: 53; saravector: 49; Sergey Kohl: 105; sevenke: 97; Shujaa_777: 23; Sidhe: 86; sma1050: 39; Son Gallery: 79; SSokolov: 101; Steve Collender: 117; steve estvanik: 151; Stocksnapper: 19; Super Prin: 119; Susan Flashman: 17; Tamara Kulikova: 143; Tom Goaz: 69; Tomasz Klejdysz: 17; treetstreet: 89; trekandshoot: 117; Victor Lauer: 89; Vladimir Wrangel: 101; Vojce: 69; wacpan: 101; Wildlife World: 143; Wlad74: 67; Yibo Wang: 143; Ysign: 37; Yvonne Baur: 85; Zerbor: 123; Zhiltsov Alexandr: 141; Zhukov Oleg: 99

The Old Design Shop: 23, 24

Wikimedia Commons: 15, 17, 24, 33, 75, 82, 101, 125, 127, 129; Biodiversity Library: 41; Internet Archive/ New York Botanical Garden: 19

www.free-scores.com: 139

All reasonable efforts have been made to trace copyright holders and to obtain their permission for the use of copyright material. The publisher apologizes for any errors or omissions in the list above and will gratefully incorporate any corrections in future reprints if notified.